THE
NONCLASSICAL ION
PROBLEM

THE NONCLASSICAL ION PROBLEM

Herbert C. Brown

Purdue University
West Lafayette, Indiana

With comments by

Paul von R. Schleyer

University of Erlangen-Nürnberg
Erlangen, West Germany

Plenum Press • New York and London

Library of Congress Cataloging in Publication Data

Brown, Herbert Charles, 1912-
   The nonclassical ion problem.

   Includes index.
   1. Carbonium ions. I. Schleyer, Paul von R., 1930-   II. Title.
QD305.C3B76   547.7   76-45175
ISBN 0-306-30950-5

©1977 Plenum Press, New York
A Division of Plenum Publishing Corporation
227 West 17th Street, New York, N.Y. 10011

Printed in the United States of America

To a fascinating, seductive intermediate,
the 2-norbornyl cation,
and to my wonderful wife,
who cheerfully coped.

Acknowledgments

The author is deeply indebted to Professor Paul von Ragué Schleyer, with whom a continuous discussion on a scientific level was maintained for many years, even though each of us occasionally arrived at different conclusions; to Dr. M. Ravindranathan, who gave exceptional assistance in preparing the drawings, checking the references and manuscript, and in assisting with the publication; and to Annette Wortman, who did her usual superlative job in typing the manuscript.

H.C.B.

The commentator has been debating the subject of nonclassical ions with Professor Brown privately for more than a decade. Our letters fill a large volume, and I am only one of the many persons with whom he has had an extensive correspondence! I must confess that many of his writings, published and unpublished, have infuriated me and have on occasion elicited intemperate responses. Nevertheless, communication has remained open and I feel that real progress in understanding the nature of carbocations has been made during this period. My own position has undergone considerable change, as has that of Professor Brown. However, we still diverge in our views of this topic. In our discussions, it is inevitable that areas of disagreement are emphasized more than areas of agreement, although the latter are large. It is very much to Professor Brown's credit that he has repeatedly attempted to arrange both open and published debates on the subject of nonclassical ions. Professor Olah has presented his viewpoint, but I have remained silent. I wish to thank Professor Brown, not only for the chance to comment on the chapters of this book, but also for his hospitality during visits to Purdue University, where most of these comments were written. The manuscript was completed during the tenure of an Adjunct Professor appointment at Case Western Reserve University, and benefited from helpful suggestions provided by Professor J. E. Nordlander and Professor G. A. Olah.

P.v.R.S.

Preface

In the early 1950s it was proposed that the cationic centers of carboniun ions in the usual solvolytic media could diminish their electron deficiency by interacting with the σ-electrons in saturated carbon–carbon or carbon–hydrogen bonds within the structure to form σ-bridges, giving new structures with markedly different properties, different symmetries, and enhanced stabilities. Such σ-bridged "nonclassical" structures were postulated to involve the formation of two electron–three center bonds of the kind present in diborane and trimethylaluminum dimer.

It was further proposed that σ-participation was important in the transition states leading to the formation of such stabilized σ-bridged cations as intermediates. Consequently, many cases of enhanced rates of solvolysis, previously attributed to relief of steric strain, were reinterpreted as involving this new phenomenon.

This new proposal of nonclassical structures for carbonium ions proved to be a popular one and was widely adopted. Indeed, in the 1950s nonclassical structures appear to have been at least considered for every carbonium ion known with the possible exception of methyl.

In science it is exceptionally rare to have a generally accepted concept challenged. Scientists become highly conservative—once a concept has become incorporated into accepted theory. This case proved to be no exception. My call for a reexamination of the concept brought forth a vigorous reaction. What became known as the "nonclassical ion controversy" has been raging for many years.

For reasons which are presented in this book, the 2-norbornyl cation became the focus of the argument. Are the high *exo* : *endo* rate and product ratios in this system the result of σ-participation in the *exo*, or are they the result of steric factors? Is this cation a symmetrical σ-bridged species in which the carbonium carbon has four nearest neighbors, with the bridging carbon (C6) having five, or do we have a pair of rapidly equilibrating classical cations with no unusual coordination numbers for the carbon atoms in the structure?

One would have thought that the application of careful experiments and intelligent thought to the problem would lead rapidly to a clear conclusion. This has not been the case. Consequently, the nonclassical ion problem provides an ideal area for examination of the actual operation of the scientific method, with its strengths and weaknesses.

Perhaps the most unfortunate consequence of the difference of opinion in this area has been the emotion it aroused. Indeed, since 1962 when I first questioned the nonclassical proposal at a Reaction Mechanisms Conference at Brookhaven, no open debate of the question was ever held. In fact, even the attempts of the editors of *Chemical and Engineering News* and of *Accounts of Chemical Research* to hold "symposia-in-print" on the subject did not materialize.

In this book I have attempted to present all pertinent data, with frank and open consideration of the pros and cons. It makes a truly fascinating story. I hope that the reader will find it as fascinating an experience to read as I encountered in pursuing this unusual hobby.

West Lafayette, Indiana   Herbert C. Brown
March 1977

Contents

List of Figures

Abbreviations

Ac	acetyl
An	anisyl
Ar	aryl
Bu	butyl
Equil. class.	equilibrating classical
Et	ethyl
eu	entropy units
eV	electron volt
glpc	gas-liquid phase chromatography
i-Pr	isopropyl
$k^{25°}$	rate constant at 25°C
k_α	polarimetric rate constant
k_c	unassisted rate constant
k_Δ	anchimerically assisted rate constant
k_s	solvent-assisted rate constant
k_t	titrimetric rate constant
Me	methyl
OBs	p-bromobenzenesulfonate
ODNB	3,5-dinitrobenzoate
OMs	methanesulfonate
ONs	p-nitrobenzenesulfonate
OPCB	p-chlorobenzoate
OPNB	p-nitrobenzoate
OTs	p-toluenesulfonate
OβNs	β-naphthalenesulfonate
p_f	partial rate factor
Ph	phenyl

RR	relative reactivity
80% acetone	80% (v/v) aqueous acetone
HFIP	hexafluoroisopropyl alcohol
HOAc	acetic acid
TFA	trifluoroacetic acid
TFE	trifluoroethanol

1

That Fascinating
Nonclassical Ion Problem

1.1. Introduction

In science it is rare that a concept, once it has been fully accepted and
incorporated into both current theory and textbooks, is successfully challenged. For fifteen years the author and his students have been engaged in a
research program directed to testing the validity of the proposal that
σ-bridges play a major role in the structures and chemistry of the 2-
norbornyl[1] and the cyclopropylcarbinyl[2] cations, as well as numerous other
σ-bridged, so-called "nonclassical ions."[3]

The problem has proved to be a truly fascinating one, difficult to resolve
to the complete satisfaction of those holding competing viewpoints. Perhaps
more important than the problem itself is the object lesson it provides on the
actual operation of the scientific method under current conditions and on
some serious deficiencies in that operation.

Accordingly, this book has five major objectives. First, to present an
objective summary of our studies and those of others directed toward
obtaining independent evidence for the proposal of σ-bridged cations.
Second, to subject to critical examination those studies which have been
published purporting to support such σ-bridged structures. (Since the great
majority of chemists active in the field were brought up on the nonclassical
ion concept, our own studies have already been subjected to such intense
critical examination.) Third, to teach students (and their professors) to
examine critically the data and arguments published even in respectable

[1]S. Winstein and D. Trifan, *J. Amer. Chem. Soc.*, **74**, 1147, 1154 (1952).
[2]J. D. Roberts and R. H. Mazur, *J. Amer. Chem. Soc.*, **73**, 3542 (1951).
[3]P. D. Bartlett, *Nonclassical Ions*, Benjamin, New York, 1965.

journals. Fourth, to stimulate others not at present committed to either viewpoint to undertake objective well-designed experiments to settle the question of the structure of the 2-norbornyl cation. Fifth, to stimulate the editors of journals to consider the desirability of achieving a fairer treatment of those holding dissenting views.

1.2. Origins

The remarkably facile rearrangements of terpenes possessing bicycloheptyl nuclei[4] has had major consequences both for the development of carbonium ion theory and for the nonclassical ion concept. Thus the relationship of the rearrangement of borneol to camphene to that of pinacolyl alcohol to tetramethylethylene was first recognized by Wagner[5] in 1899. It was Meerwein's study of the rearrangement of camphene hydrochloride (**1**) to isobornyl chloride (**2**) (1) that resulted in the first suggestion

$$ (1) $$

of carbonium ions (**3, 4**) as intermediates in such a molecular transformation[6] (2). Such Wagner–Meerwein rearrangements are still fascinating organic chemists.

$$ (2) $$

In 1939 it was suggested by Wilson and his coworkers that such a rapidly equilibrating pair of cations (**3** ⇌ **4**) might exist instead as the mesomeric species[7] **5**. Previously **5** would have been considered to be the

[4] J. Simonsen and L. N. Owens, *The Terpenes*, Vol. II, Cambridge University Press, London, 1949.
[5] G. Wagner and W. Brickner, *Ber.*, **32**, 2302 (1899).
[6] H. Meerwein and K. van Emster. *Ber.*, **55**, 2500 (1922).
[7] T. P. Nevell, E. de Salas, and C. L. Wilson, *J. Chem. Soc.*, 1188 (1939).

transition state separating **3** and **4**. In effect, Wilson proposed that this transition state might be sufficiently stable so as to become a minimum in the reaction path, doing away with the need to consider the classical structures, **3** and **4**.

5

1.3. The Nonclassical Ion Era

Very little attention appears to have been paid to Wilson's suggestion for some ten years.[7] However, in the early 1950s the study of carbonium ions became very active.[8] σ-Bridged structures were proposed for the 2-norbornyl[1] and cyclopropylcarbinyl[2] cations. These caught the fancy of the chemical public. The concept was widely adopted and used. Indeed, it would appear that nonclassical structures were at least considered for every known aliphatic, alicyclic, and bicyclic carbonium ion,[3] with the possible exception of the methyl cation. Several representative systems are shown in Figure 1.1.

The question to be resolved in systems, such as are shown in Figure 1.1, is whether the carbonium ion intermediate generated in the reaction under consideration exists as the static classical ion (top structure of each system), or as a rapidly equilibrating pair (or set) of cations (bottom structures of each system), with a transition state as shown, or whether this transition state represents a stable intermediate, one so stable that the classical structures need no longer be considered.

The enthusiasm for such σ-bridged formulations was great and the usual scientific caution was often ignored. To illustrate the tenor of the times, two statements made at the Symposium on Carbonium Ions at the Meeting of the American Chemical Society in St. Louis in 1961 may be cited.[9] One speaker, questioned as to why he proposed a particular nonclassical structure for the cation under discussion, replied, "Because it is fashionable." Another speaker, asked the same question for another species, replied, "Because it looks so nice."

It is clear that inadequate heed was being paid to the counsel of R. B. Woodward that the mere fact that one is dealing with fugitive intermediates should not convey license to propose highly fanciful structures without adequate experimental support.[10]

[8] A. Streitwieser, Jr., *Solvolytic Displacement Reactions*, McGraw-Hill, New York, 1962.
[9] H. C. Brown, *Chem. Eng. News*, **45**, 87 (Feb. 13, 1967).
[10] R. B. Woodward, *Perspectives in Organic Chemistry*, A. Todd, Ed., Interscience, New York, 1956, pp. 177–178.

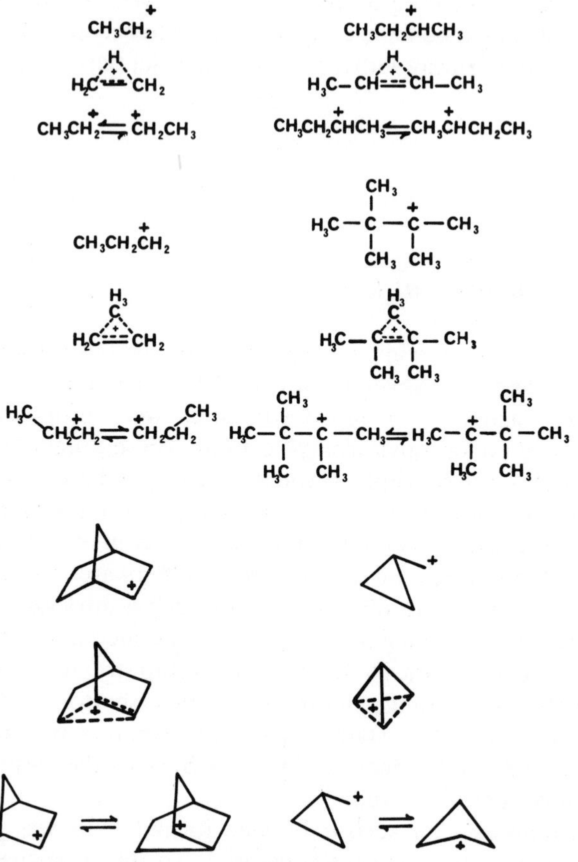

Figure 1.1. Representative carbonium ions for which σ-bridged nonclassical structures have been considered.

1.4. Steric Assistance

For certain of the objectives of this book, it might be of interest to point out how I became interested in the question. As I have pointed out elsewhere,[11] it has been my experience that one research problem leads to another. The exploration of discrepancies in research results often opens up new research areas. Differences between one's own interpretation of chemical phenomena and those of others also lead into new areas. The latter was the case here.

[11]H. C. Brown, *Boranes in Organic Chemistry*, Cornell University Press, Ithaca, N.Y., 1972.

Our studies of the factors influencing the stabilities of molecular addition compounds had led me to the conclusion that three bulky alkyl groups attached to nitrogen must constitute a center of strain.[11] If so, accumulation of three alkyl groups around a carbon atom must likewise constitute a center of strain. Consequently, I suggested in 1946 that such derivatives should exhibit enhanced rates of solvolysis arising from the relief of steric strain as the crowded initial state passes through the transition state to the less crowded, less strained planar carbonium ion[12] (3).

$$
\begin{array}{ccc}
\overset{\displaystyle R}{\underset{\displaystyle R}{R\!-\!\!\!-\!C\!-\!Cl}} \;\rightarrow\; & \overset{\displaystyle R}{\underset{R\;\;\;\;R}{C^{+}}} & +\,Cl^{-} \qquad\qquad (3)\\[2mm]
\text{tetrahedral} & \text{planar} & \\
\text{(strained)} & \text{(less strained)} &
\end{array}
$$

We explored the proposal in a number of representative systems.[13,14] The results supported the proposed interpretation. The concept appeared to be generally accepted.[15] Accordingly, we left this area for other problems.

1.5. An Alternative Interpretation

I soon began to observe that the new concept of σ-bridging was being used with increasing frequency to account for fast rates of solvolysis in systems where relief of steric strain appeared to provide an adequate interpretation. For example, Ingold and his coworkers attributed the high rate of solvolysis of camphene hydrochloride (1) (6000 times that of *tert*-butyl chloride) not to the relief of steric strain as the chloride ion departs its sterically crowded environment (1), but to the driving force provided by the formation of a stabilized σ-bridged cation (5)[16] (4).

$$
\mathbf{1} \qquad\qquad \mathbf{5}
$$

Similarly, Bartlett suggested that the formation of a methyl-bridged tri-*tert*-butylcarbinyl cation (7) might provide the driving force to account for

[12]H. C. Brown, *Science*, **103**, 385 (1946).
[13]H. C. Brown and R. S. Fletcher, *J. Amer. Chem. Soc.*, **71**, 1845 (1949).
[14]H. C. Brown and H. L. Berneis, *J. Amer. Chem. Soc.*, **75**, 10 (1953).
[15]E. L. Eliel, *Steric Effects in Organic Chemistry*, M. S. Newman, Ed., Wiley, New York, 1956, Chapter I.
[16]F. Brown, E. D. Hughes, C. K. Ingold, and J. F. Smith, *Nature*, **168**, 65 (1951).

the observed high rates of solvolysis of tri-*tert*-butylcarbinyl derivatives[17] (**6**)
(**5**).

$$
\begin{array}{ccc}
& \overset{\displaystyle C(CH_3)_3}{\underset{\displaystyle C(CH_3)_3}{|}} & \overset{\displaystyle C(CH_3)_3}{|} \\
(CH_3)_3C-\overset{|}{C}-X & \rightarrow & (CH_3)_3-C \\
& & \quad\;\; \overset{}{\underset{(CH_3)_2C}{}} \; \overset{+}{\cdot}CH_3 \;+\; X^- \\
& \mathbf{6} & \mathbf{7}
\end{array}
\tag{5}
$$

Likewise, Heck and Prelog suggested that the high rate of solvolysis of cyclodecyl tosylate (**8**) might be due to the driving force associated with the formation of a stabilized bridged cyclodecyl cation[18] (**9**), rather than to the decrease in internal strain (6) (Section 2.5).

$$\mathbf{8} \qquad\qquad\qquad \mathbf{9} \tag{6}$$

Originally, I had no reason to question these proposals. However, it was interesting that the phenomenon appeared to be significant only for structures where my coworkers and I had anticipated that relief of steric strain would be an important factor in enhanced rates of solvolysis. Accordingly, I began to examine in detail the available data for such systems in order to clarify how much of the rate enhancement might be due to relief of steric strain and how much to this new phenomenon of bridging by saturated carbon. We then undertook experimental studies of selected systems.

These studies persuaded me that there was no convincing evidence for σ-bridging in these systems. All of the observed rate enhancements could be accounted for without including this new feature.

1.6. The Rococo Period of Carbonium Ion Structures

Meanwhile, the theory was undergoing rapid elaboration. Many new structures were being proposed and adopted, some of exceptional complexity. This might well be termed the "rococo period" of carbonium ion structures (Figure 1.2).

At this point, I was faced with one of those agonizing decisions. The nonclassical ion concept had proved to be exceedingly popular. A large fraction of the physical organic chemists in the United States were engaged in research in this area. The subject had entered the textbooks and

---

[17]P. D. Bartlett, *J. Chem. Educ.*, **30**, 22 (1953).
[18]R. Heck and V. Prelog, *Helv. Chim. Acta*, **38**, 1541 (1955).

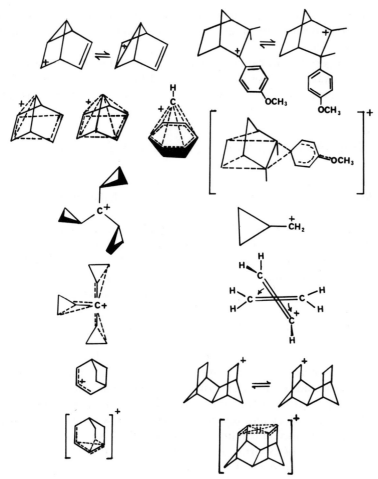

Figure 1.2. Representative carbonium ions from the "rococo period" of carbonium ion structures.

thousands of graduate students were being taught and examined annually on the subject matter and approved interpretation. Obviously, to challenge the theory would lead to severe repercussions. However, to remain silent might subject future generations of students to needless studies of erroneous concepts.

There comes a time when one cannot remain on the sidelines. It becomes necessary to stand forth and declare, "But the Emperor is naked!"

Accordingly, in March 1961, at the Symposium on Carbonium Ions held at the St. Louis Meeting of the American Chemical Society, I pointed out that, in my opinion, many of the proposed nonclassical structures for carbonium ions rested on exceedingly fragile experimental foundations, and I suggested the desirability of further experimental work to test and reaffirm these foundations.

I had hoped that my cautious comments[19] would set in motion a critical objective reexamination of the field. Instead they were apprehended as a heresy, triggering what appeared to be a "holy war" to prove me wrong.

1.7. Difficulties in Challenging an Accepted Theory[11]

For the special objectives of this book, it appears appropriate to consider some of the problems encountered in challenging an accepted theory.

Many theoretical proposals are advanced and subjected to experimental test. The uncovering of unfavorable experimental data results in the ready revision of the proposal or its withdrawal. The situation is quite different for a theory that has reached the stage of wide acceptance. Such a theory appears in the textbooks, is taught to a new generation of students, and becomes accepted as a fixed part of the fabric of established chemical theory. It becomes exceedingly difficult to question such an accepted concept.

In an ideal operation of the scientific method, all of the experiments carried out involving a particular theory would appear in print. An unsatisfactory theory leads to many experimental results that will not confirm the theory. As these accumulate, there will be general recognition that the theory is unsatisfactory and revision of the theory will be carried out. In actual fact, under present conditions, this is not the case—there takes place an actual selection of experimental results to favor an established theory.

Let us consider an example. A young man receives a post in a university. He considers possible research problems. He has just been taught the fascinating new nonclassical ion theory, and it appears to be an interesting area to explore. He makes a prediction based on the theory and subjects it to experimental test. The nonclassical theory is qualitative; consequently, even on a purely statistical basis, the chances are $50:50$ that the results will support his prediction.

If this turns out to be the case, he is happy as a lark. The study is written up and submitted to a journal. The editor, reading what appears to be an interesting study supporting an accepted theory, sends it off to two experts for review. These will be supporters of the theory, since nonsupporters of an

[19]H. C. Brown, "The Transition State," *Chem. Soc. (London) Spec. Publ.*, **16**, 140–158, 174–178 (1962).

accepted theory will be rare (and under suspicion). These supporters of the nonclassical paper will read it, find it interesting, and return it with a recommendation that it be accepted. Shortly it will appear in print.

However, suppose that the experimental result proves to be contrary to the prediction. The young man will be puzzled. Here is a generally accepted theory that he has been taught, yet the results do not fit. The chances are that he will decide to put the study on the shelf for a while, to permit it to "mature," and often it will remain there forever.

Suppose that the young man is a rare individual who decides that the contrary result should be published. He submits the paper. The editor sees that a young unknown worker is proposing to publish results contrary to a well-established theory. He immediately sends it to two of the leading workers in the field, strong supporters of the theory. These referees ask, "Who is this young man who dares to attack such a well-established theory?" Clearly there must be something wrong with his experiments. They return the paper with the recommendation that the manuscript not be published until the author has done additional experiments, and they outline sufficient additional work to take years to complete.

Consequently, there occurs a steady selection of papers to support an established theory.

1.8. Further Difficulties—A "Soft" Theory

The nonclassical ion theory is a qualitative theory—a "soft" theory. It can be adjusted to accommodate almost any result. Let us consider several examples.

The *exo*:*endo* rate ratio for 2-norbornyl brosylate is 350.[1] What *exo*:*endo* rate ratio should we predict for *exo*-5,6-trimethylene-2-norbornyl derivatives (7)? Suppose the *exo*:*endo* rate ratio, **10/11**, turned

$$\tag{7}$$

TsO — **10** / OTs — **11**

out to be 10,000, vastly greater than that in the parent system. This increased rate ratio can be interpreted to support the nonclassical ion interpretation. The proponents would argue that the additional strain causes the 1,6-bond to be more polarizable and therefore better able to contribute to the displacement of the *exo*-tosyl group. However, the *exo*:*endo* rate ratio is actually decreased to 11.2.[20] Now it can be argued that the decreased rate ratio supports the nonclassical interpretation because the additional ring inhibits σ-participation!

The introduction of *gem*-dimethyl substituents (**13**) into the 6-position of 2-norbornyl (**12**) markedly reduces the rate of solvolysis[21] (8). It was argued that the *gem*-dimethyl substituents would be expected to interfere sterically with σ-bridging from the 6- to the 2-carbon.

(8)

	12	**13**
RR (25°C):	280 (*endo* = 1.00)	21.6

On the other hand, the introduction of *gem*-dimethyl groups (**15**) in the 5-position of 2-bicyclo[2.1.1]hexyl tosylate (**14**) markedly increases the rate of solvolysis[22] (9). It is now argued that the tertiary bridge is better able to stabilize the electron-deficient center in the transition state leading to the nonclassical ion **17** than the corresponding primary bridge in **16**[22] (10)!

(9)

	14	**15**
RR (75°C):	1.00	36

(10)

Originally, it was reported that there was a reduced α deuterium effect ($k_H/k_D = 1.10$) in *exo*-norbornyl, but a normal effect ($k_H/k_D = 1.20$) in *endo*-norbornyl[23] (11). This reduced α deuterium effect was considered to support the nonclassical delocalized intermediate. More recently, it has been reported that the low α-deuterium effect was the result of scrambling of the deuterium substituent in the *exo* isomer.[24] The revised value, 1.21, is

[20]K. Takeuchi, T. Oshika, and Y. Koga, *Bull. Chem. Soc. Japan*, **38**, 1318 (1965).

[21]P. v. R. Schleyer, M. M. Donaldson, and W. E. Watts, *J. Amer. Chem. Soc.*, **87**, 375 (1965).

[22]J. Meinwald, Abstracts, 18th National Organic Chemistry Symposium, Columbus, Ohio, 1963, pp. 37–44; J. K. Crandall, Ph.D. Thesis, Cornell University, Ithaca, N.Y., 1965.

[23]C. C. Lee and E. W. C. Wong, *J. Amer. Chem. Soc.*, **86**, 2752 (1964).

[24]D. E. Sunko and S. Borčić, *Isotope Effects in Chemical Reactions*, C. J. Collins and N. S. Bowman, Eds., Van Nostrand–Reinhold, New York, 1970, Chapter 3.

essentially identical for the two isomers. But now the identical value for the secondary isotope effects is considered to support the nonclassical formulation![24]

$$RR \ (25°C): \qquad 1.21 \qquad\qquad 1.11 \qquad\qquad (11)$$

Olah has supported the σ-bridged structure for the cyclopropylcarbinyl cation for the reason that the ^{13}C shift calculated for a set of equilibrating cations does not agree with the observed value.[25] To make this calculation, he must first assume a value for the ^{13}C shift for a carbon atom of a primary cation. Unfortunately, it is not possible to test his conclusion by showing agreement between the calculated ^{13}C shift and that calculated for the nonclassical structure—no one has yet claimed that such shifts can be calculated for such structures.

Yet it has been such "soft" data and arguments which provide the heart of the arguments for the nonclassical formulation.

1.9. Still Further Difficulties—Selective Reviews

The great majority of workers in the carbonium ion field have either contributed to the development of the nonclassical interpretation or have been taught by those who have. Consequently, it is natural that reviews of the nonclassical ion problem are overwhelmingly written by those who favor the nonclassical ion interpretation.[3,8,26-29] The mere weight of numbers tends to be persuasive.

One means of overcoming this handicap would be to hold open discussions where the conflicting viewpoints could be discussed before an uncommitted objective audience. However, just as in the political field where it is difficult to get an incumbent to hold an open discussion with a challenger, so it has proved impossible in recent years to arrange an open discussion of this subject. Indeed, even the attempts by the editors of

[25]G. A. Olah, C. L. Jeuell, D. P. Kelly, and R. D. Porter, *J. Amer. Chem. Soc.*, **94**, 147 (1972).

[26]J. A. Berson, *Molecular Rearrangements*, P. de Mayo, Ed., Wiley-Interscience, New York, 1963, Chapter 3.

[27]G. D. Sargent, *Carbonium Ions*, Vol. III, G. A. Olah and P. v. R. Schleyer, Eds., Wiley, New York, 1972, Chapter 24.

[28]G. A. Olah, *Carbocations and Electrophilic Reactions*, Wiley, New York, 1974.

[29]W. J. le Noble, *Highlights of Organic Chemistry*, M. Dekker, New York, 1974.

Chemical and Engineering News and of *Accounts of Chemical Research* to organize "symposia-in-print" on the subject failed.[30] Perhaps the format adopted for the present volume will serve to remedy this deficiency.

1.10. Conclusion

It is obviously not easy for one individual with one research group to question such a well-established theory. Obviously it is not practical to make an intensive reexamination of each of the thousands of papers in the literature purporting to support the theory. While he is occupied in demonstrating that a particular study is ambiguous and can be reinterpreted, another hundred papers supporting the theory will appear.

I decided that the only feasible strategy was to select two systems generally believed to be the best possible examples for nonclassical ions: the cyclopropylcarbinyl and the 2-norbornyl cations. These systems were then subjected to intensive reexamination. The results of these studies were then reported in communication form, in order to avoid engaging in long, unprofitable polemical discussions prior to the time when I had reached a definite decision. I believe the time is now ripe for such a decision.

---

[30]See Editor's Note in H. C. Brown, *Accounts Chem. Res.*, **6**, 377 (1973).

Comments

In science, general principles are more important than individual examples. While cations of the 2-norbornyl type have played an historic role, the bonding illustrated by such nonclassical ions is of far greater significance. G. N. Lewis emphasized in 1916[31] that the majority of molecules are well described by bonds assigned two electrons each. Although exceptions were known even at that time, e.g., benzene, Lewis' ideas were so compelling that generations of organic chemists have subsequently been trained to believe that carbon is capable of being bonded only to a maximum of four adjacent atoms. The bridged 2-norbornyl cation and other nonclassical ions violate this principle. The traditional view point was expressed by Brown in 1967, "... ions such as the bicyclobutonium and the norbornyl in its σ-bridged form ... do not possess sufficient electrons to provide a pair for all the bonds required by the proposed structures. A new bonding concept not yet established in carbon structures is required."[32]

It is indeed difficult to challenge an accepted theory and to establish a "new bonding principle." Even today, organic chemists are puzzled when they encounter the numerous stable uncharged molecules in which carbon is clearly pentacoordinate, hexacoordinate, heptacoordinate, and even octacoordinate. Even though there are now *hundreds* of such examples, many with precise structures secured by diffraction or microwave methods, these embarrassing compounds are almost never mentioned in "organic" textbooks, but are relegated to the "inorganic" world. Molecules of this type, metal alkyls, carboranes, carbide metallic carbonyl clusters, etc., share with carbocations the common feature that they are "electron deficient." Such molecules do not possess sufficient electrons to simultaneously satisfy the octet rule for all atoms and to provide two electrons for each chemical bond. Consequently, electron deficient molecules tend to adopt structures which are unusual from the conventional organic chemist's viewpoint. Nevertheless, three-center two-electron bonds and multicenter bonding are, in fact, now completely established "in carbon structures."

I think it is important for organic chemists to appreciate more of the world of carbon compounds which have been developing rapidly outside their normal sphere of interest, a world where bridged structures are the rule, rather than the exception. Let us examine a few of these compounds whose structures are secured by X-ray or other methods. Afterwards, the bridging exhibited by nonclassical carbocations will appear to be unexceptional, if not inevitable.

[31]G. N. Lewis, *J. Amer. Chem. Soc.*, **38**, 762 (1916).
[32]H. C. Brown, *Chem. Eng. News*, **45**, No. 7, 87 (1967).

Carbide Carbonyl Clusters[33]

An increasing list of molecules are known in which a carbon atom is surrounded by a number of transition metal atoms which, in turn, have attached carbonyls. Rather than tinkling around loosely in these metallic cages, the carbons contribute importantly to the bonding[34] holding these ensembles together (Figure 1.3).

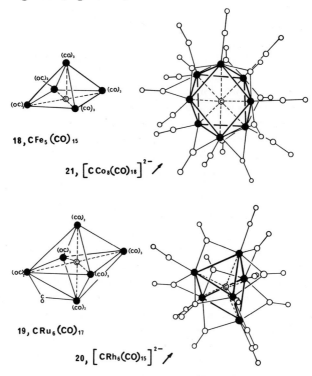

Figure 1.3. Carbide carbonyl clusters with multicoordinate carbon.

In **18**,[35] the central carbon is surrounded in C_{4v} symmetry by five iron atoms, and is slightly below the center of the Fe_4 ring. Since the Fe–C distances are all roughly the same, the conclusion that the central carbon is pentacoordinate is inescapable.

[33]Review: P. Chini, G. Longoni, and V. G. Albano, *Advan. Organometallic Chem.*, **14**, 285 (1976).

[34]K. Wade, *Chemistry in Britain*, 177 (1975); *Advan. Inorg. Chem. Radiochem.*, **18**, 1 (1976).

[35]E. H. Braye, L. F. Dahl, W. Hubel, and D. L. Wampler, *J. Amer. Chem. Soc.*, **84**, 4633 (1962).

Hexacoordinate carbon is present in **19**[36] and **20**.[37a] The former possesses approximate octahedral (O_h) symmetry, with the six bonds to carbon extending at right angles to one another along the axes of the p_x, p_y, and p_z atomic orbitals. Additional examples of octahedral type are known.[37,38] Structure **20** possesses a trigonal prism of six rhodium atoms with the D_{3h} carbon in the center.[37a]

The highest known coordination number of carbon is eight. This is found in beryllium carbide, where each carbon is surrounded by a cube of beryllium atoms,[39] in **21**,[37b] where carbon is at the center of a tetragonal antiprism, and possibly in $COs_8(CO)_{21}$.[40]

Carboranes[34,41]

The carboranes are carbon molecules with remarkable structures, bonding, and chemical properties deserving the attention of organic chemists. Boron, like nitrogen, is adjacent to carbon in the periodic table, and, due to the efforts of Professor Brown, is already employed widely in organic synthesis. Carboranes often are astonishingly stable thermally and chemically. Furthermore, they undergo electrophilic substitution just like benzene, and there are excellent grounds for considering these three-dimensional molecules to be "aromatic."

The skeletal bonds of the carboranes are often partial and possess less than two electrons each. However, the atoms involved have coordination numbers greater than four, and the greater number of bonds compensates for the weaker nature of each bond. The structures of these species follow simple rules, and are easy to understand from a molecular-orbital

[36] (a) B. F. G. Johnson, R. D. Johnston, and J. Lewis, *J. Chem. Soc. A*, 2865 (1968); (b) A. Sirigu, M. Bianchi, and E. Benedetti, *Chem. Commun.*, 596 (1969).

[37] (a) V. G. Albano, M. Sansoni, P. Chini, and S. Martinengo, *J. Chem. Soc. Dalton*, 651 (1973). Also see V. G. Albano, P. Chini, S. Martinengo, M. Sansoni, and D. Strumolo, *Chem. Commun.*, 299 (1974); (b) V. G. Albano, P. Ciani, M. Sansoni, D. Strumolo, B. T. Heaton, and S. Martinengo, *J. Amer. Chem. Soc.*, **98**, 5027 (1976).

[38] E.g., $CFe_6(CO)_{16}^{2-}$, M. R. Churchill, J. Wormald, J. Knight, and M. J. Mays, *J. Amer. Chem. Soc.*, **93**, 3073 (1971). $CRu_6(CO)_{14}$ (arene), Ref. 36a and R. Mason and W. Robinson, *Chem. Commun.*, 468 (1968); and $CCo_6(CO)_{14}^{2-}$, Ref. 37a.

[39] F. A. Cotton and G. Wilkinson, "Advanced Inorganic Chemistry," 3rd ed., Wiley-Interscience, New York, 1972, p. 291.

[40] C. R. Eady, B. F. G. Johnson, and J. Lewis, *J. Chem. Soc. Dalton*, 2606 (1975).

[41] There are numerous books and reviews, *Inter alia*: (a) W. N. Lipscomb, "Boron Hydrides," W. A. Benjamin, Inc., New York, 1963; (b) T. Onak, "Organoborane Chemistry," Academic Press, New York, 1975; (c) E. L. Muetterties, Ed., "Boron Hydride Chemistry," Academic Press, New York, 1975; (d) R. E. Williams, *Advan. Inorg. Chem. Radiochem.*, **18**, 67 (1976).

viewpoint.[34,41] It suffices here to present only a very limited selection of examples. These are arranged below in order of increasing carbon coordination number (Figure 1.4). It should be noted that each carbon bears a hydrogen, and is situated above rings bearing four or five boron or boron and carbon atoms. For references to the structures of the individual species, the reader is referred to the reviews cited.[41]

PENTACOORDINATE CARBONS

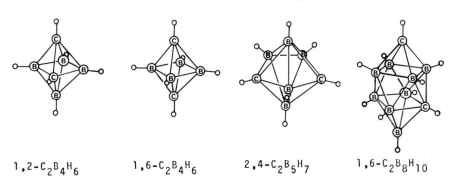

$1,2-C_2B_4H_6$ $1,6-C_2B_4H_6$ $2,4-C_2B_5H_7$ $1,6-C_2B_8H_{10}$

HEXACOORDINATE CARBONS

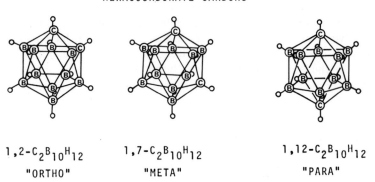

$1,2-C_2B_{10}H_{12}$ $1,7-C_2B_{10}H_{12}$ $1,12-C_2B_{10}H_{12}$
"ORTHO"   "META"   "PARA"

Figure 1.4. Carboranes containing multicoordinate carbons.

Pyramidal Carbocations[42]

"Rococo" carbocations are still very much with us. The pyramidal ions **22** and **23**, which have been studied theoretically, are the parents of

[42]Review: H. Hogeveen and P. W. Kwant, *Accounts Chem. Res.*, **8**, 413 (1975). See Section 14.5.

substituted systems for which experimental evidence has been reported.[42] While this evidence seems most secure for the hexamethyl derivative (**24**) of Hogeveen, the case for these ions is very much strengthened by the existence of well-defined analogies in carboranes and boranes.[34,41,42] Thus, **22** is isoelectronic with **25** and **26**, **23** with **27–31**. The Hogeveen dication **24** has an exact counterpart in neutral **32** (Figure 1.5).[43]

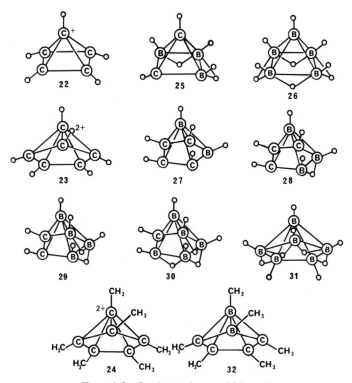

Figure 1.5. Isoelectronic pyramidal species.

Ions **22–24** thus provide a bridge of understanding between the "organic" and "inorganic" worlds; the simple molecular-orbital descriptions of **22**[44] and **23**[42] can be extrapolated to virtually all of the carboranes and boranes.[45] If these systems are regarded as being constituted of CH or BH "caps" interacting with rings, *six* electrons are needed to fill the three stabilized orbitals resulting from the interactions. For this reason, **33** is not

[43]J. Hasse, *Z. Naturforsch.*, **28a**, 785 (1973).
[44]W.-D. Stohrer and R. Hoffmann, *J. Amer. Chem. Soc.*, **94**, 1661 (1972).
[45]P. v. R. Schleyer and associates, unpublished calculations.

stable,[46] since it possesses *eight* such electrons (six from benzene and two from the $:C^+H$ "cap"); the extra two electrons in **33** have to occupy an unfavourable, high energy orbital. Structures like **33** should be possible if they possess two fewer valence electrons; **34**[42] is unrealistic electrostatically, but our calculations[45] indicate **35** to be very stable. Structure **36** appears to be an excellent candidate for a neutral molecule of this type with a heptacoordinate bridging carbon.

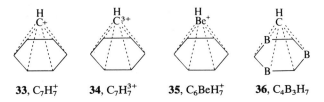

33, $C_7H_7^+$ **34**, $C_7H_7^{3+}$ **35**, $C_6BeH_7^+$ **36**, $C_4B_3H_7$

Summary

Professor Brown has made it very clear to me that he does not question the three-center two-electron bonding involved in, e.g., the bridged 2-norbornyl cation on *theoretical* grounds. We agree that the bridged and the classical structures cannot lie very far apart in energy, although it is probable that only one form is an energy minimum. I take the view that since three center and multicenter bonding involving carbon is so widespread in electron deficient molecules, the probability of bridging in carbocations must be high. Many nonclassical carbocations have been postulated in the literature, some erroneously, and some without adequate basis. However, many bridged ions appear to me to exist, and a purpose of this book is to examine the available evidence from two different viewpoints.

Positively charged carbocations are highly electron deficient, more so than analogous neutral boron derivatives. Resonance stabilization, hyperconjugation, and bridging are ways such electron deficiency can be stabilized internally, but interaction with the environment (solvation) may often be more effective. Depending on the degree and nature of the solvation, the energy, the detailed geometry, and perhaps even the gross structure may be altered.

In discussing the nonclassical problem, the enormous range of possibilities and conditions have to be kept in mind. We will be considering evidence in the gas phase, in superacid media, in organic solvents like TFA (trifluoroacetic acid) with low nucleophilicity, and finally in solvents like

[46]C. Cone, M. J. S. Dewar, and D. Landman, *J. Amer. Chem. Soc.*, **99**, 372 (1977).

acetic acid and aqueous acetone with high nucleophilicity. The nature and behavior of a given species may be quite different under such widely varying conditions.

Nevertheless, many carbocations are relatively stable and will tend to be relatively unaffected by changes in environment. Other species, like CH_3^+, $C_2H_5^+$, and CH_5^+, are so unstable that direct observation in condensed phases is impossible; only highly encumbered "carbonoid" or "cationoid" species can be expected. The norbornyl cation is intermediate in stability, and the detailed mechanism of solvolysis, as well as the problem of solvation, is a very real one when considering this species.

The present emphasis on the problem of the structure of nonclassical ions should not be allowed to detract from the established importance of σ-electron delocalization in organic chemistry. Winstein and others showed beyond doubt that neighboring σ-bonds could participate during solvolytic generation of carbocations (see comments, Chapter 3). Olah has developed a whole new chemistry involving electrophilic substitution of saturated hydrocarbons.[28] Whether bridged ions are involved as intermediates or "only" as transitions states in the rich variety of reactions he has discovered is less important than the demonstration that C—C and C—H single bonds are not "inert," but function as electron donors under suitable conditions.

Steric Assistance in Solvolytic Processes

2.1. Introduction

The dissociation of molecular addition compounds provided a powerful new tool to explore steric effects in organic molecules.[1,2] To illustrate the application of this tool, the behavior of trimethylamine and triethylamine with trimethylborane may be reviewed. Triethylamine (pKa 10.75) is a somewhat stronger base than trimethylamine (pKa 9.75). Consequently, one would anticipate that the stronger base would form the more stable addition compound with the Lewis acid. However, the reverse is true: $\Delta H_{dissoc.}(CH_3)_3N:B(CH_3)_3$, 17.6 kcal mol^{-1}; $\Delta H_{dissoc.}(C_2H_5)_3N:B(CH_3)_3$, ~10 kcal mol^{-1}.

It is possible to account for this apparent anomaly in terms of the large steric requirements of the three ethyl groups of triethylamine in the addition compound (**2**) (1). In this structure it is possible to rotate only two of the

$$\Delta H_{dissoc.} \text{ kcal mol}^{-1} \quad 17.6 \qquad\qquad \sim 10 \qquad\qquad\qquad 20.0$$

[1]H. C. Brown, *J. Chem. Ed.*, **36**, 424 (1959).
[2]H. C. Brown, *Boranes in Organic Chemistry*, Cornell University Press, Ithaca, N.Y., 1972, Chapters V–VIII.

three ethyl groups out of the space required by the trimethylborane molecule—the third must project out so as to interfere with the adding molecule (**2**).

This interpretation was tested with quinuclidine,[3] a base of strength (pKa 10.95), comparable to triethylamine. In this molecule the three ethyl groups of triethylamine are in effect rotated to the rear of the molecule and effectively held there by a methine carbon away from the trimethylborane moiety (**3**). Indeed, the addition compound formed by trimethylborane and quinuclidine is far more stable, $\Delta H_{dissoc.}$ 20.0 kcal mol^{-1}, even more stable than the compound formed by trimethylamine (**1**).

Such studies led to the conclusion that three bulky alkyl groups attached to nitrogen constitute a center of strain. It followed then that three alkyl groups attached to carbon must also constitute a center of strain. Accordingly, it was suggested that such derivatives should exhibit enhanced rates of solvolysis arising from relief of steric strain as the crowded initial state passes through a less crowded transition state on the way to the less strained planar carbonium ion[4] (**2**).

$$R-\overset{\displaystyle R}{\underset{\displaystyle R}{C}}-Cl \rightarrow \overset{\displaystyle R}{\underset{\displaystyle R \quad R}{C^+}} + Cl^- \qquad (2)$$

tetrahedral   planar
(strained)   (less strained)

2.2. Steric Assistance in the Solvolysis of Highly Branched Alkyl Derivatives

This proposal was explored.[5,6] In a number of systems significant rate accelerations with increasing steric requirements of the alkyl groups attached to the carbinyl carbon were observed (**3**).

RR (25°C):   1.00   21   580   (3)

[3]H. C. Brown and S. Sujishi, *J. Amer. Chem. Soc.*, **70**, 2878 (1948).
[4]H. C. Brown, *Science*, **103**, 385 (1948).
[5]H. C. Brown and R. S. Fletcher, *J. Amer. Chem. Soc.*, **70**, 1845 (1948).
[6]H. C. Brown and H. L. Berneis, *J. Amer. Chem. Soc.*, **75**, 10 (1953).

Numerous related results have been recorded. For example, Bartlett and Tidwell observed that trimethyl-, trineopentyl-, and tri-*tert*-butylcarbinyl *p*-nitrobenzoates undergo solvolysis at relative rates of 1.00, 560, and 13,000,[7] respectively (4).

RR (40°C):   1.00   13,000   (4)

Relief of steric strain can result in enormous rate effects. For example, the *cis,cis,cis*-perhydro-9-phenalenyl *p*-nitrobenzoate (4) undergoes hydrolysis at a rate some 10^6 times faster than the *cis,cis,trans*-stereoisomer (5).[8]

RR (25°C):   4,000,000   1.00

2.3. Steric Assistance in the Relative Effects of Methyl and tert-Butyl Groups

Some time ago it was suggested that many of the unusual characteristics of the norbornyl system may have their origin in unusually large steric strains arising from the rigidity of this bicyclic structure.[9] It was pointed out that strains arising from the presence of a bulky substituent would be small in the relatively flexible aliphatic system, larger in the less flexible alicyclic system, and enormous in the rigid bicyclic norbornyl system. This proposal was tested by examining the relative effects of methyl and *tert*-butyl substituents upon the rates of solvolysis of a selected series of derivatives.[10]

[7]P. D. Bartlett and T. T. Tidwell, *J. Amer. Chem. Soc.*, **90**, 4421 (1968).
[8]H. C. Brown and W. C. Dickason, *J. Amer. Chem. Soc.*, **91**, 1226 (1969).
[9]H. C. Brown and J. Muzzio, *J. Amer. Chem. Soc.*, **88**, 2811 (1966).
[10]E. N. Peters and H. C. Brown, *J. Amer. Chem. Soc.*, **97**, 2892 (1975).

The rate of solvolysis of *tert*-butyldimethylcarbinyl *p*-nitrobenzoate (**7**) is faster than that of *tert*-butyl *p*-nitrobenzoate (**6**) by a factor of 4.4 (6). Thus in this aliphatic system, the replacement of a methyl group by the *tert*-butyl group increases the rate by a relatively small factor, presumably a manifestation of the relief of steric strain. It should be recalled that the accumulation of two or three bulky groups at the tertiary center results in far larger rate enhancements[7] (4).

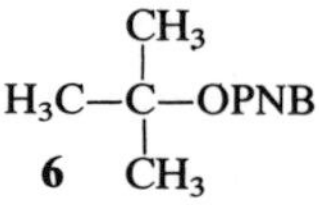

$$H_3C-\underset{\underset{\textbf{6}\ CH_3}{|}}{\overset{\overset{CH_3}{|}}{C}}-OPNB \qquad\qquad H_3C-\underset{\underset{\textbf{7}\ H_3C\ CH_3}{|}}{\overset{\overset{H_3C\ CH_3}{|\ \ \ |}}{C}-C}-OPNB \qquad (6)$$

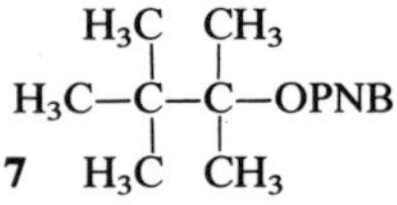

RR (25°C):   1.0   4.36

Replacement of the 1-methyl group in 1-methylcyclopentyl *p*-nitrobenzoate (**8**) by the more bulky *tert*-butyl group (**9**) results in a rate enhancement by a factor of 112 (7).

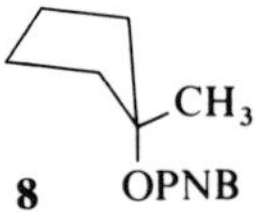

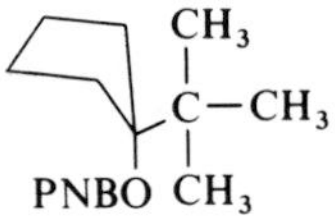

(7)

RR (25°C):   1.00   112

Similarly, the replacement of the 1-methyl group in 1-methylcyclohexyl *p*-nitrobenzoate (**10**) by the *tert*-butyl group (**11**) increases the rate of solvolysis by a similar factor, 134 (8).

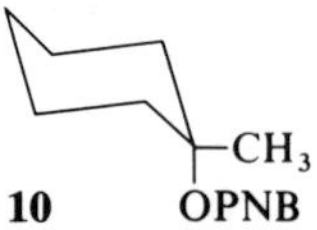

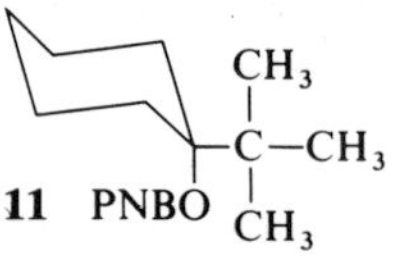

(8)

RR (25°C):   1.00   134

In the cycloheptyl system the effect of *tert*-butyl (**13**)/methyl (**12**) is even larger, 273 (9).

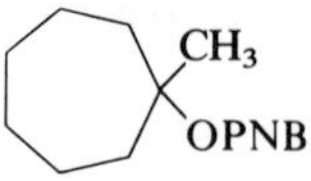

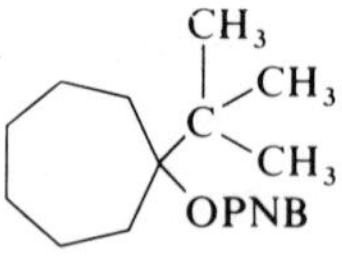

(9)

RR (25°C):   1.00   273

In the relatively rigid bicyclics, the effect of replacing methyl by *tert*-butyl is considerably greater.

In 3-nortricyclyl (**15/14**) the effect is modest, apparently a consequence of the reduced steric requirements of the ring system resulting from the formation of the cyclopropane ring (10).

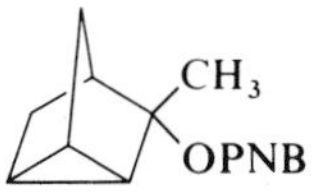

(10)

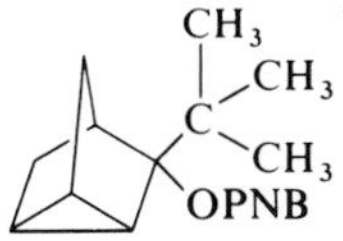

	14	**15**
RR (25°C):	1.00	1790

The effect is much larger in the related *endo*-norbornyl derivatives (11), rising to 39,600 for *tert*-butyl (**17**)/methyl (**16**).

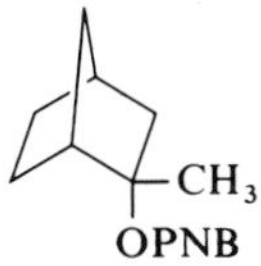

(11)

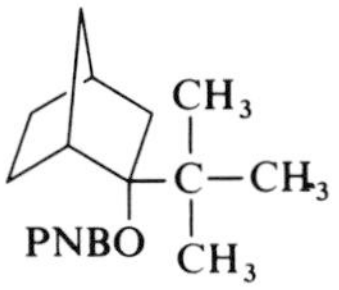

	16	**17**
RR (25°C):	1.00	39,600

The 2-adamantyl *p*-nitrobenzoates[11] (12) exhibit an even larger rate enhancement for *tert*-butyl (**19**)/methyl (**18**).

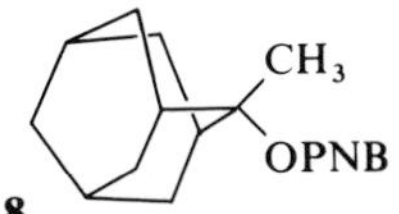

(12)

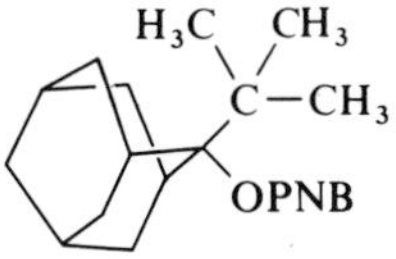

	18	**19**
RR (25°C):	1.00	225,000

Finally, the introduction of *gem*-dimethyl groups into the 3-position of the norbornyl structure greatly increases the strain and raises the effect of *tert*-butyl (**21**)/methyl (**20**) to 1,120,000 (13).

These results are summarized in Table 2.1.

(13)

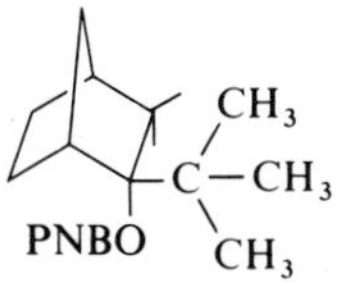

	20	**21**
RR (25°C):	1.00	1,120,000

[11]J. L. Fry, E. M. Engler, and P. v. R. Schleyer, *J. Amer. Chem. Soc.*, **94**, 4628 (1972).

TABLE 2.1. Data for the Effects of Methyl and tert-Butyl on the Rates of Solvolysis of Tertiary p-Nitrobenzoates[a,b]

p-Nitrobenzoate	$10^6 k_1$, sec^{-1c} at 25.0°C	$\dfrac{k(t\text{-Bu})}{k(\text{Me})}$	$\Delta H^\ddagger$, kcal mol^{-1}	$\Delta S^\ddagger$, eu
tert-Butyl	7.45×10^{-5}	4.36	29.2	−7.1
tert-Butyldimethylcarbinyl	3.25×10^{-4}		29.0	−4.8
1-Methyl-1-cyclopentyl	2.11×10^{-3}	112	26.9	−7.9
1-tert-Butyl-1-cyclopentyl	0.236		24.6	−6.5
1-Methyl-1-cyclohexyl	5.48×10^{-5}	134	30.1	−4.4
1-tert-Butyl-1-cyclohexyl	7.35×10^{-3}		28.5	−0.1
1-Methyl-1-cycloheptyl	4.21×10^{-3}	273	27.1	−6.0
1-tert-Butyl-1-cycloheptyl	1.15		24.6	−3.1
3-Methyl-3-nortricyclyl	1.81×10^{-3}	1,790	26.4	−9.1
3-tert-Butyl-3-nortricyclyl	3.24			
2-Methyl-endo-norbornyl	1.13×10^{-5}	39,600	30.2	−7.5
2-tert-Butyl-endo-norbornyl	0.48		25.4	−2.4
9-Methyl-9-bicyclo[3.3.1]-nonyl	3.34×10^{-4}	104,000	28.4	−6.5
9-tert-Butyl-9-bicyclo-[3.3.1]nonyl	34.8			
2-Methyl-2-adamantyl	1.43×10^{-4}	225,000	30.2	−2.2
2-tert-Butyl-2-adamantyl	34.2		21.6	−6.5
2-Methyl-endo-camphenilyl	2.31×10^{-5}	1,120,000	30.4	−5.1
2-tert-Butyl-endo-camphenilyl				

[a] In 80% acetone.
[b] Reference 10.
[c] Calculated from data at higher temperatures.

In the past it has not been uncommon to estimate steric interactions in norbornyl structures from their magnitudes in aliphatic and, especially, alicyclic systems.[12] The A factors determined for alicyclic systems have been an especially fertile source for such estimates. The present results reveal that such estimates can be seriously in error. Steric effects in norbornyl derivatives can be huge compared with the effects we are accustomed to dealing with in the more flexible aliphatic and alicyclic derivatives.

2.4. Steric Effects in Norbornyl Derivatives

As will be discussed later, the high exo:endo rate ratio in 2-norbornyl derivatives appears to have its origins in steric effects (Chapter 8). The

[12] G. D. Sargent, Quart. Revs., 20, 301 (1966).

U-shaped structure of the norbornyl system offers steric resistance to the departure of the *endo* leaving group.

If the *exo:endo* rate ratio is the result of steric forces, it should be possible to vary these rates in a predictable manner by varying the steric environment. For example, the *exo:endo* rate ratio for the solvolysis of the 2-methylnorbornyl *p*-nitrobenzoates (**22, 23**) in 80% aqueous acetone is

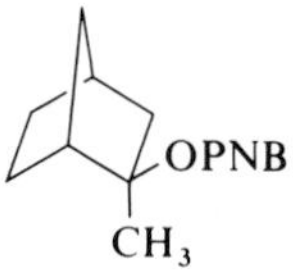

(14)

| RR (25°C): | 1.00 | 885 |

885^{13} (14). The introduction of *gem*-dimethyl substituents into the 7-position of **22** and **23**, as in the 2,7,7-trimethyl-2-norbornyl *p*-nitrobenzoates (**24, 25**), should result in an increase in the steric crowding of the *exo* environment of the norbornyl system, while maintaining the *endo* environment essentially constant. If the steric interpretation of the *exo:endo* ratio is correct, such a modification should result in a greatly decreased *exo:endo* rate ratio.[14] This is observed (15).

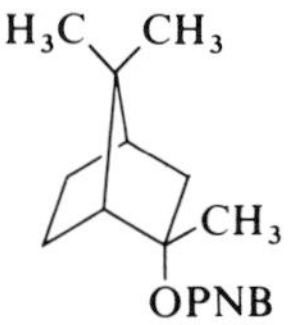

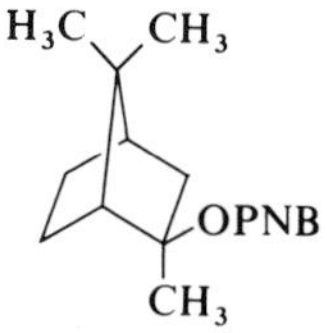

(15)

| RR (25°C): | 1.00 | 6.1 |

On the other hand, an increase in the steric crowding of the *endo* environment (**26**), while maintaining the *exo* environment essentially constant (**27**), as in the 2,6,6-trimethyl-2-norbornyl *p*-nitrobenzoates (16), should result in an increase in the *exo:endo* rate ratio. This is observed[13] (16).

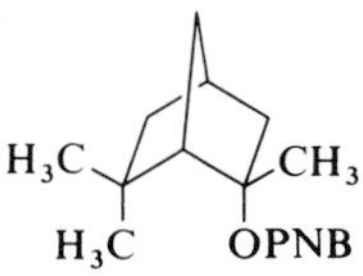

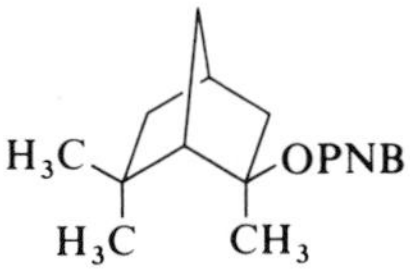

(16)

| RR (25°C): | 1.00 | 3,630,000 |

The data are summarized in Table 2.2.

[13]S. Ikegami, D. L. Vander Jagt, and H. C. Brown, *J. Amer. Chem. Soc.*, **90**, 7124 (1968).
[14]H. C. Brown and S. Ikegami, *J. Amer. Chem. Soc.*, **90**, 7122 (1968).

TABLE 2.2. *Rates of Solvolysis of 2,7,7-Trimethyl-2-Norbornyl and 2,6,6-Trimethyl-2-Norbornyl p-Nitrobenzoates and Related Derivatives*[a,b]

p-Nitrobenzoate	$10^6 k_1$, sec^{-1c} at 25°C	Relative rate	$\Delta H^{\ddagger}$, kcal mol^{-1}	$\Delta S^{\ddagger}$, eu
1-Methylcyclopentyl	2.11×10^{-3}	1.00	26.9	−7.9
2-Methyl-*exo*-norbornyl	1.00×10^{-2}	4.74	26.3	−7.0
2-Methyl-*endo*-norbornyl	1.13×10^{-5}	0.00536	30.2	−7.5
2,7,7-Trimethyl-*exo*-norbornyl	4.01×10^{-2}	19.0	24.5	−10.2
2,7,7-Trimethyl-*endo*-norbornyl	6.54×10^{-3}	3.1	25.8	−9.5
2,6,6-Trimethyl-*exo*-norbornyl	7.26	3440	23.2	−4.1
2,6,6-Trimethyl-*endo*-norbornyl	2.00×10^{-6}	0.000948	31.5	−6.4

[a] In 80% aqueous acetone.
[b] References 13, 14.
[c] Calculated from data at higher temperatures.

Consequently, by increasing the steric crowding of the *exo* face, the *exo*:*endo* ratio decreases from 885 in the parent compound to 6.1. By increasing the steric crowding of the *endo* face, the *exo* : *endo* ratio increases from 885 in the parent compound to 3,630,000. Surely steric effects must play a major role in the high *exo*:*endo* rate ratios exhibited by 2-norbornyl derivatives.

2.5. Steric Effects in Ring Systems

In a masterful discussion of the chemistry of ring systems, Prelog revealed that chemical behavior varied in a fairly consistent manner with ring size.[15] It was of interest to see if the concepts which had been previously used with considerable success to rationalize the chemistry of 5- and 6-membered ring systems[16] could be extended to larger rings. Accordingly, we undertook to examine the rates of solvolysis of the 1-methyl-1-chlorocycloalkanes,[17] the acetolysis of the cycloalkyl tosylates,[18,19] and the rates of reaction of sodium borohydride with the cyclanones.[20]

It was pointed out previously that reactions involving a change in coordination number from four to three are favored in the 5-ring, presumably because the change is accompanied by a decrease in internal strain.[16]

[15] V. Prelog, *J. Chem. Soc.*, 420 (1950).
[16] H. C. Brown, J. H. Brewster, and H. Shechter, *J. Amer. Chem. Soc.*, **76**, 467 (1954).
[17] H. C. Brown and M. Borkowski, *J. Amer. Chem. Soc.*, **74**, 1894 (1952).
[18] R. Heck and V. Prelog, *Helv. Chim. Acta*, **38**, 1541 (1955).
[19] H. C. Brown and G. Ham, *J. Amer. Chem. Soc.*, **78**, 2735 (1956).
[20] H. C. Brown and K. Ichikawa, *Tetrahedron*, **1**, 221 (1957).

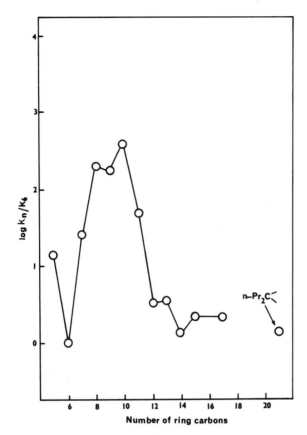

Figure 2.1. The effect of ring size on the rate of acetolysis of the cycloalkyl tosylates at 70°C.

This reaction is unfavorable in the relatively strain-free 6-ring system. The 7-ring is more strained and reactions of this kind are favored. The strain increases to a maximum at the 10-ring.[15] We should expect to find the rates for reactions of this kind, involving a change in coordination number of a ring atom from four to three, to reach a maximum at this point. Indeed, the rate of acetolysis of the cyclodecyl tosylate does reveal a maximum with the 10-ring (Figure 2.1).

Contrariwise, the reverse should be true for reactions involving a change in coordination number from three to four, as in the reaction of the cycloalkanones with sodium borohydride. Indeed, a minimum in rate is observed for cyclodecanone (Figure 2.2).

Linear free-energy relationships involving aliphatic derivatives generally are not observed. Consequently, it is of interest that the ring compounds from 5- through 10-ring members provide a reasonably good linear free-

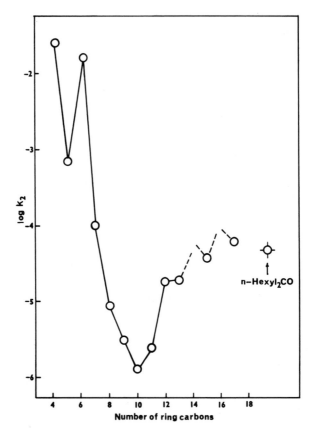

Figure 2.2. The effect of ring size on the rate of reaction of the cyclanones with sodium borohydride in isopropyl alcohol at 0°C.

energy relationship between the rates of reaction of acetolysis of the tosylates and the rates of reaction of the cycloalkanones with borohydride (Figure 2.3). (The relationship fails with the larger, more flexible rings.[20])

2.6. Conclusion

These results clearly establish that relief of steric strain is an important factor in the rates of solvolysis of sterically strained systems where such strain can be relieved in the transition state leading to the carbonium ion. The question now to be considered is whether σ-bridging is an important factor in such fast rates. The rate maximum observed in cyclodecyl tosylate[18,19] corresponds to a rate minimum for cyclodecanone.[20] These

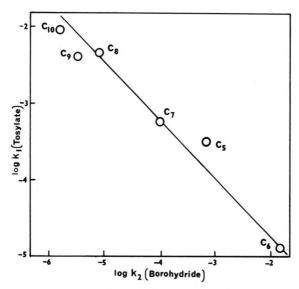

Figure 2.3. Free-energy relationship between the rates of acetolysis of the cycloalkyl tosylates and the rates of reaction of the cyclanones with sodium borohydride.

phenomena appear to have a common origin (Figure 2.3). Is it reasonable to attribute the maximum in cyclodecyl tosylate to σ-bridging,[18] an explanation that cannot explain the apparently related slow rate in cyclodecanone?

Comments

Appreciation of the important role of steric effects is one of H. C. Brown's major contributions to chemistry. Crowded molecules are higher in energy; if their strain is relieved in the reaction transition states, accelerated rates will result. This is abundantly demonstrated in the solvolysis of many tertiary systems which display rate enhancements of startlingly large magnitude.

However, serious problems are involved in the application of steric theory. Such theory is also "soft" since there is no way at present to estimate qualitatively the effect on solvolysis rate produced by steric strains. As the following pair of examples illustrate, it is not even possible to predict reliably whether a solvolysis will be accelerated or decelerated by increased crowding! It is evident from a comparison of **22** and **26** that strain involving the leaving group is not always relieved on ionization (17).

$$\text{decelerated by } gem\text{-dimethyl} \tag{17}$$

	22 OPNB	**26** CH$_3$ OPNB
RR(25°C):	1.0	0.18
(75°C):	1.0	0.24

However, the rate enhancement observed in (18) indicates that it is not a simple matter to predict where such steric effects will involve an increase or decrease in rate.

$$\text{accelerated by } gem\text{-dimethyl} \tag{18}$$

RR(75°C)[21]:	1.0	14

Evidently, the locus of departure of the leaving group is different in (17) and (18). What is needed is a simple and accurate method to calculate energy differences between ground and transition states.

The application of steric theory to secondary (and to primary) systems, where most of the work on σ-electron participation has been carried out, is even more problematical. Such systems are less crowded and steric effects are expected to play a more minor role. Corresponding secondary and

---

[21]D. Faulkner, M. A. McKervey, D. Lenoir, C. A. Senkler, and P. v. R. Schleyer, *Tetrahedron Lett.*, 705 (1973).

tertiary substrates often respond oppositely to the introduction of substituents:

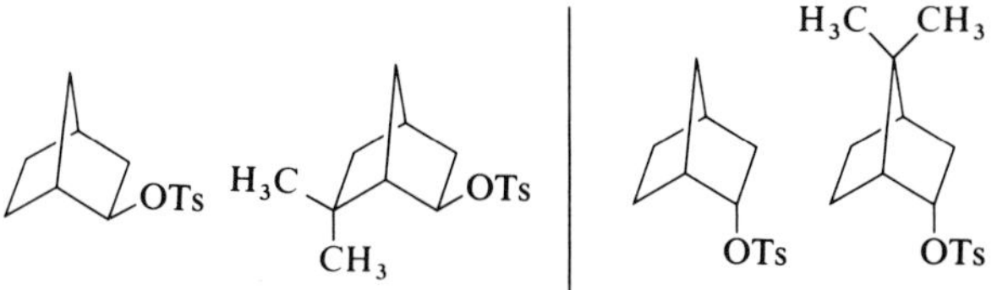

RR (HOAc, 75°C): 1.0	0.07	1.0	0.8

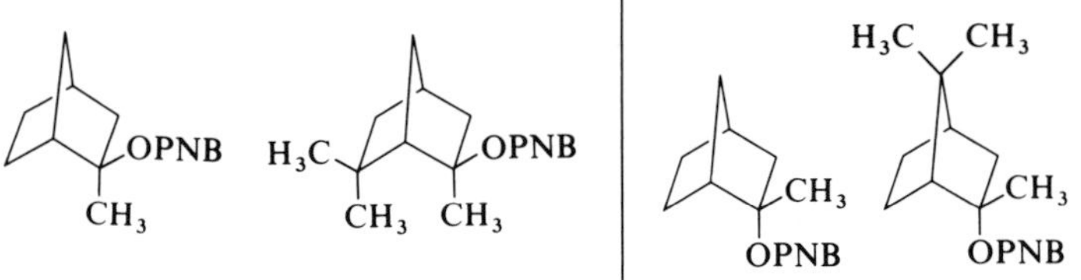

RR (75°C): 1.0	338	1.0	197
(25°C): 1.0	726	1.0	578

Brown's interpretation of the behavior of these tertiary derivatives is quite reasonable. The question is the extent to which we can carry over such interpretations to the secondary systems. The behavior illustrated above appears to go beyond mere differences in the steric requirements of methyl vs. hydrogen. For this reason, it does not seem possible to me to use tertiary norbornyl derivatives to model the behavior of seemingly analogous secondary systems. It has *not* been demonstrated that steric effects "play a major role in the high *exo*: *endo* ratios exhibited by *secondary* 2-norbornyl derivatives."

The quantitative prediction of solvolysis rates remains a challenging goal of physical organic chemistry, although progress has been made.[22] Unfortunately, linear free-energy approaches to this problem, such as that suggested by Figure 2.3, are not general, as Brown himself appreciated.[20] When $\log k_1$ (tosylate) is compared with $\log k_2$ (borohydride) for a larger number of systems, no correlation is observed.[23]

---

[22]R. C. Bingham and P. v. R. Schleyer, *J. Amer. Chem. Soc.*, **93**, 3189 (1971); W. Parker, R. L. Tranter, C. I. F. Watt, L. W. K. Chang, and P. v. R. Schleyer, *ibid.*, **96**, 7121 (1974).

[23]P. v. R. Schleyer, private communication to H. C. Brown; W. L. Dilling, C. E. Reineke, and R. A. Plepys, *J. Org. Chem.*, **34**, 2605 (1969).

3

σ-Participation—A Factor
in Fast Rates?

3.1. Introduction

The results discussed in Chapter 2 clearly establish the importance of relief of steric strain in the fast rates of solvolysis exhibited by a number of sterically crowded systems. However, as was pointed out in Chapter 1, the development of the concept of σ-bridging in the early 1950s and its growing utilization to reinterpret the structures of carbonium ions led to increasing numbers of proposals attributing enhanced rates in such systems not to relief of steric strain but to the stabilization provided by the σ-bridge.

3.2. n-, π-, and σ-Participation

It may be helpful to review the relationship of the proposed carbon participation (π- and σ-) to neighboring group phenomena.[1,2]

The brilliant studies of Winstein and his coworkers in the 1940s established that donor atoms (those containing available n electrons) in the β-position could greatly enhance the rates of solvolytic reactions and simultaneously control the mechanism (and stereochemistry) of substitution (1).

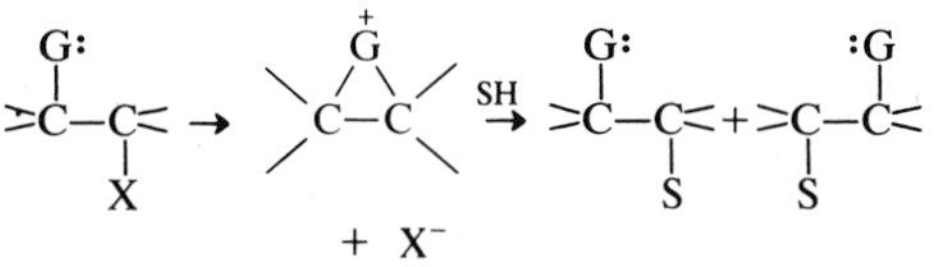

$$\tag{1}$$

[1]A. Streitwieser, Jr., *Solvolytic Displacement Reactions*, McGraw-Hill, New York, 1962.
[2]B. Capon, *Quart. Rev. (London)*, **18**, 45 (1964).

Originally, those working in the field believed that a large rate enhancement was essential in order to postulate the formation of a stabilized bridged intermediate. For example, the four-fold rate enhancement observed for *trans*-2-chlorocyclohexyl brosylate, compared with the *cis* isomer, was not considered significant by Winstein and his coworkers, and the formation of a chloronium intermediate was not proposed[3] (2). However, the large factors of 800 for the bromine neighboring group and 10^6 for the iodine derivative were believed to reflect the driving force accompanying the formation of the bridged intermediates.

RR (75°C):	4	800	2,700,000	(2)

Consequently, even with the relatively favorable class of *n*-donors, there exists a wide variation in their ability to serve as such bridging groups. Groups such as thioalkoxy and iodo participate strongly. On the other hand, there are groups such as methoxy and chloro which participate very weakly or not at all (Table 3.1).

The π-subdivision of neighboring groups also exhibits a wide range of effectiveness[2] (3).

RR (25°C):	2.1	160	100,000,000	(3)

Consequently, within these two donor classes there is a wide variation in the ability to participate, varying from strong participation to very weak or negligible participation. The question we face is whether a saturated carbon–carbon bond, the σ-classification, can serve as a donor to a developing carbonium ion center.

The rate of solvolysis of *anti*-7-norbornenyl tosylate[4] (2) is faster than that of the saturated derivative (1) by a factor of 10^{11} (4). Clearly the

---

[3] S. Winstein, E. Grunwald, and L. L. Ingraham, *J. Amer. Chem. Soc.*, **70**, 821 (1948); E. Grunwald, *J. Amer. Chem. Soc.*, **73**, 5458 (1951).

[4] S. Winstein, M. Shatavsky, C. Norton, and R. B. Woodward, *J. Amer. Chem. Soc.*, **77**, 4183 (1955).

TABLE 3.1. *Classification of Neighboring Groups*

n-	π-	σ-
RS		CH₃
R₂N		
I		
Br		
Cl		
RO		

reaction is proceeding with participation of the double bond to give a cationic intermediate whose structure will be discussed later (Section 4.6).

	1	**2**	(4)
RR (50°C):	1.00	10^{11}	

On the other hand, the related structure (**4**) undergoes solvolysis slower, not faster than the saturated derivative (**3**)[5] (**5**). Clearly π-participation is not significant here. It is evidently a sensitive function of the structure and cannot be assumed to occur automatically—it must be demonstrated in each case.

	3	**4**	(5)
RR (50°C):	1.00	1/8.3	

[5] S. Winstein and J. Sonnenberg, *J. Amer. Chem. Soc.*, **83**, 3235 (1961).

It is evident that σ-participation must be much more difficult. Consider the solvolysis of 7-norbornyl tosylate. The exceptional slowness of the rate provides no reason to postulate σ-participation by the 2,3-carbon bond (**6**)

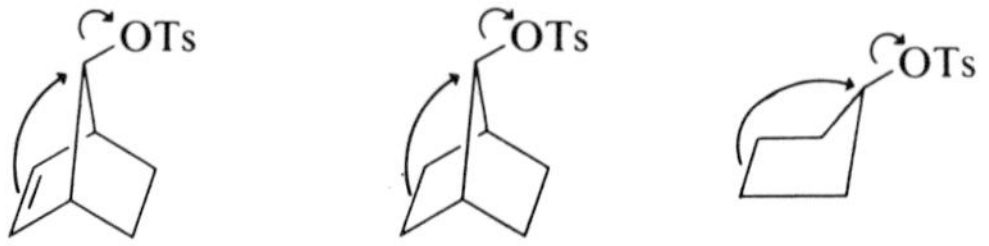

(6)

in the manner evident in *anti*-7-norbornenyl (**5**) (6). Such σ-participation appears to be absent also in the related cyclopentyl derivative (**7**). If it occurred, it would lead to a σ-bridged nonclassical ion, with a three orbital-two electron bond, involving both the developing orbital of C7 and the bonding orbitals of the C2–C3 bond (**8**).

One possible modification is to have the developing orbital on C7 interact with only one of the two carbons of the ethano bridge.

3.3. σ-Participation and Fast Rates

As was pointed out earlier (Section 1.5), the fast rate of solvolysis of camphene hydrochloride (**9**), 6000 times greater than the rate of *tert*-butyl chloride, was attributed to the driving force provided by the formation of a σ-bridged cation (**10**)[6] (7).

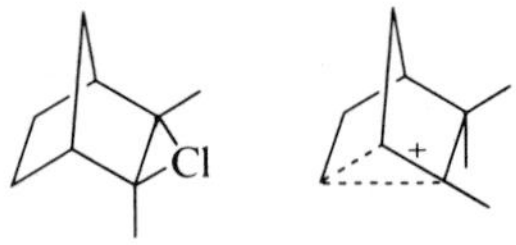

(7)

Likewise, the fast rate of solvolysis of tri-*tert*-butylcarbinyl *p*-nitrobenzoate (**11**), 13,000 times that for *tert*-butyl *p*-nitrobenzoate, was attributed not to relief of steric strain but to the driving force associated with the formation of a σ-bridged intermediate (**12**)[7] (8).

[6]F. Brown, E. D. Hughes, C. K. Ingold, and J. F. Smith, *Nature*, **168**, 65 (1951).
[7]P. D. Bartlett, *J. Chem. Ed.*, **30**, 22 (1953).

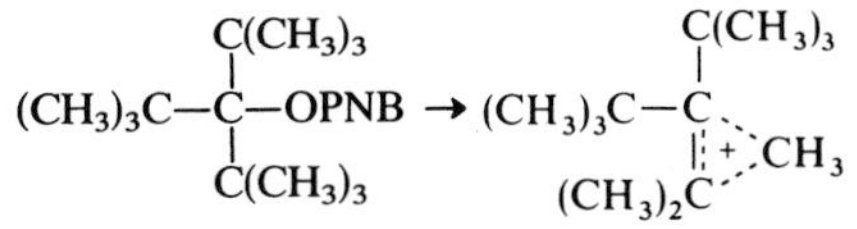

$$(CH_3)_3C-\overset{\displaystyle C(CH_3)_3}{\underset{\displaystyle C(CH_3)_3}{\underset{|}{\overset{|}{C}}}}-OPNB \rightarrow (CH_3)_3C-\overset{\displaystyle C(CH_3)_3}{\underset{(CH_3)_2C}{\overset{|}{C}}}\overset{+}{\cdots}CH_3 \tag{8}$$

11   **12**

Similarly, the fast rate of acetolysis of cyclodecyl tosylate (**13**) was attributed not to relief of nonbonded interactions, but to the driving force provided by the formation of a stabilized σ-bridged species (**14**)[8] (9).

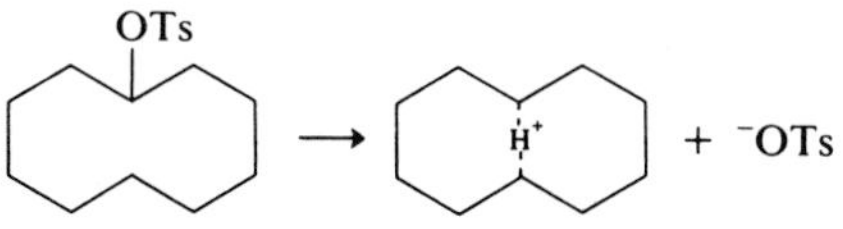

$$\text{(9)}$$

13   **14**

It appeared to be a remarkable coincidence that all such cases of enhanced rates of solvolysis attributed to σ-participation involved strained molecules, where relief of steric strain might be considered a major contributing factor. It may be appropriate to point out the essential difference in the two proposals.

3.4. σ-Participation vs. Steric Assistance

The solvolysis of *tert*-cumyl chloride (**16**) in ethanol proceeds at a rate some 4600 times greater than that of *tert*-butyl chloride (**15**) (10).

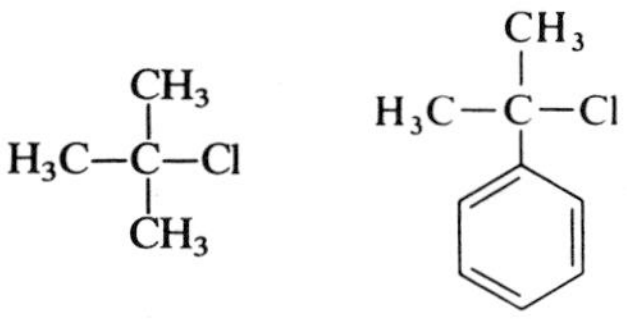

$$\text{(10)}$$

	15	**16**
RR (25°C):	1.00	4600

The enhanced rate of solvolysis of *tert*-cumyl chloride is attributed to resonance stabilization by the phenyl group of the incipient carbonium ion in the transition state. Such resonance stabilization is greater than the hyperconjugative stabilization provided by the corresponding methyl group. Consequently, the enhanced rate is primarily the result of the existence of a transition state which is lower in energy than the corresponding transition state in *tert*-butyl chloride (Figure 3.1).

[8]R. Heck and V. Prelog, *Helv. Chim. Acta*, **38**, 1541 (1955).

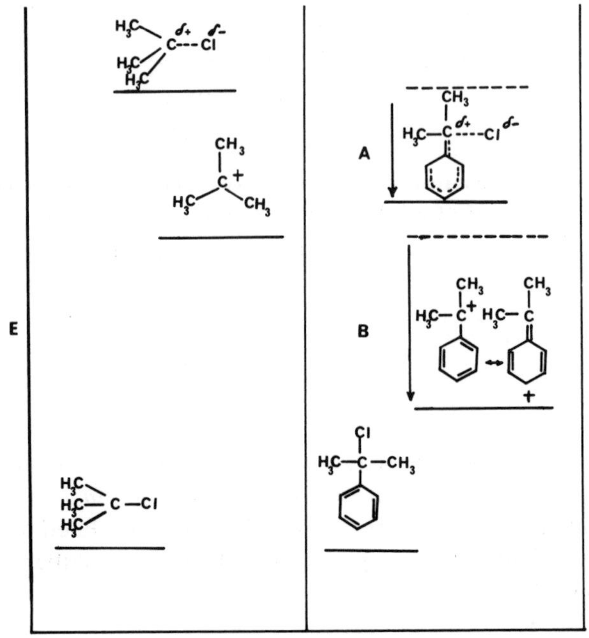

Figure 3.1. Effect of classical resonance in facilitating solvolysis through stabilization of the transition state A and the carbonium ion intermediate B.

It was proposed that an increase in the steric requirements of the three alkyl groups of the tertiary chloride could result in steric strain, increasing the ground state energy. Since such strain should be decreased in the transition state leading to the planar carbonium ion, solvolysis of such derivatives should be facilitated (Figure 3.2). The competing proposal was that σ-bridging could occur in the transition state, stabilizing that state and facilitating solvolysis (Figure 3.2).

It is interesting to examine what has happened to these three proposals for enhanced rates attributed to σ-participation.

3.5. Steric Assistance—Not σ-Participation

As was mentioned earlier, the magnitude of the enhanced rate of solvolysis exhibited by camphene hydrochloride (6000) as compared to *tert*-butyl chloride (1.00) was considered by Ingold to be not compatible with relief of steric strain.[6] Consequently, the enhanced rate was attributed to the driving force associated with the formation of a stabilized mesomeric cation (**10**).

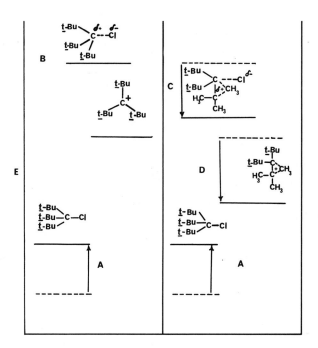

Figure 3.2. Effect of steric strain in the ground state A facilitating solvolysis either by relief of strain in the less-strained transition state B or by stabilization via σ-bridging in the transition state C and in the carbonium ion intermediate D.

In such analyses it is important to select suitable models. A question may be raised as to the suitability of *tert*-butyl chloride as a model for the highly sterically congested molecule, camphene hydrochloride (**17**). A more suitable model would doubtless be the pentamethylcyclopentyl chloride (**18**) realized by opening the 5,6-ethylene bridge of **17** (11).

Regrettably, this structure has not yet been synthesized. However, many of the other methyl-substituted 1-methylcyclopentyl chlorides have been prepared and their rates of ethanolysis at 25°C determined.[9] The data

[9]H. C. Brown and F. J. Chloupek, *J. Amer. Chem. Soc.*, **85**, 2322 (1963).

TABLE 3.2. *Rates of Ethanolyses at 25°C of Camphene Hydrochloride, 2-Methyl-exo-norbornyl Chloride and Related Methyl-Substituted 1-Methylcyclopentyl Chlorides*

Chloride	$10^6 k_1$, sec^{-1} at 25°C	Relative rates			
tert-Butyl	0.085	1.00			
2,3,3-Trimethyl-exo-norbornyl	1,160	13,600^a	206	5.7	
2-Methyl-exo-norbornyl	30.2	355	5.4		2.1
1-Methylcyclopentyl	5.62	66	1.00		
1,2-Dimethylcyclopentyl	14.5^b	171	2.6		1.00
	10.6^c	125	1.9		
1,2,2,5-Tetramethylcyclopentyl	202	2,380	36	1.00	
1,2,2,5,5-Pentamethylcyclopentyl	458	5,390	82		

^{a}The rate constant at 0°C, 37.2×10^{-6} sec^{-1}, is in reasonable agreement with the earlier value (Reference 6). The discrepancy with the relative rate reported earlier (6000) must arise from the rate constant for tert-butyl chloride, which was not given explicitly.
^{b}Chloride from cis-1,2-dimethylcyclopentanol.
^{c}Chloride from trans-1,2-dimethylcyclopentanol.

(Table 3.2) do not support the conclusion that the rates for the norbornyl derivatives are exceptional (12).

$$13,600 \qquad\qquad 2380$$

$$H_3C-\underset{\underset{CH_3}{|}}{\overset{\overset{CH_3}{|}}{C}}-Cl \qquad\qquad (12)$$

$$1.00$$

$$355 \qquad\qquad 66$$

Quite clearly the results are in accord with the postulated effect of increasing steric strain in enhancing the rates of solvolysis of highly branched tertiary chlorides (Section 2.2). Similar results have been realized in other comparisons.[10]

[10]H. C. Brown, F. J. Chloupek, and M.-H. Rei, *J. Amer. Chem. Soc.*, **86**, 1246, 1247, 1248 (1964).

These results do not disprove the possible formation of mesomeric norbornyl cations in the solvolysis of secondary 2-norbornyl derivatives. However, they do eliminate the original argument that the rates of solvolysis of camphene hydrochloride and other *exo*-norbornyl derivatives are too fast to be explicable on any basis other than the formation of mesomerically stabilized cations.

Solvolysis of optically active 1,2-dimethyl-*exo*-norbornyl *p*-nitrobenzoate (**19**; X = OPNB) in 90% (v/v) aqueous acetone gives alcohol with 9% retention.[11] Similarly, methanolysis of optically active 1,2-dimethyl-*exo*-norbornyl chloride (**19**; X = Cl) gives methyl ether with 14% retention.[12] Goering and Clevenger conclude that they are trapping a rapidly equilibrating classical ion (**20**) or ion pair before it has completely

$$(13)$$

equilibrated[12] (13). The *σ*-bridged species (**21**) possesses a plane of symmetry and would be optically inactive.

If one can extrapolate these results for a tertiary 2-norbornyl cation to the related 2,3,3-trimethyl-2-norbornyl cation (**22**) from camphene hydrochloride (**17**), then this cation must also be a classical species without a *σ*-bridge (14).

$$(14)$$

Investigation has also removed the other two proposals for *σ*-bridging as a basis for fast rates (8) and (9). In the case of cyclodecyl tosylate, it was

[11]H. Goering and K. Humski, *J. Amer. Chem. Soc.*, **90**, 6213 (1968).
[12]H. L. Goering and J. V. Clevenger, *J. Amer. Chem. Soc.*, **94**, 1010 (1972).

observed that the introduction of transannular deuterium failed to affect the rate significantly (9). Consequently, the σ-bridged formulation (14) was withdrawn.[13,14] A detailed study of the products from the solvolysis of tri-*tert*-butylcarbinyl *p*-nitrobenzoate (8) failed to support the σ-bridged formulation (12) and this has been withdrawn.[15]

3.6. The Fast Rates of Exo-Norbornyl and Cyclopropylcarbinyl

As discussed earlier, my interest in the proposals for σ-participation as a basis for fast rates arose from the fact that this proposal provided an alternative to relief of steric strain in accounting for such fast rates in sterically strained derivatives. Now that the original proposals had been eliminated, I might well have considered the problem solved and withdrawn from the field.

However, I had become interested in the general question of σ-bridging and nonclassical ions. It was generally agreed that 2-norbornyl[16] and cyclopropylcarbinyl[17] provided the most favorable examples for such σ-participation leading to the formation of σ-bridged (nonclassical) cations. For example, the rate of solvolysis of *exo*-norbornyl brosylate is greater than that of the *endo* isomer by a factor of 350.[16] This was attributed to σ-participation in the *exo* isomer (23) leading to the formation of a

$$\text{23} \qquad\qquad \text{24} \qquad\qquad\qquad\qquad (15)$$

σ-bridged (nonclassical) cation (24) (15). Similarly, the rate of solvolysis of cyclopropylcarbinyl derivatives (25) is exceptionally fast. This again was attributed to σ-participation by one or both of the far carbon atoms of the cyclopropyl ring to give a σ-bridged (nonclassical) cation (26) (16).

$$\text{25} \qquad\qquad \text{26} \qquad\qquad\qquad\qquad (16)$$

Accordingly, our program was expanded to include these systems (Chapters 5 and 6).

[13]V. Prelog, *Rec. Chem. Prog.*, **18**, 247 (1957).
[14]V. Prelog and J. G. Traynham, *Molecular Rearrangements*, Vol. 1, P. de Mayo, Ed., Interscience, New York, 1963, Chapter 9.
[15]P. D. Bartlett and T. T. Tidwell, *J. Amer. Chem. Soc.*, **90**, 4421 (1968).
[16]S. Winstein and D. Trifan, *J. Amer. Chem. Soc.*, **74**, 1147, 1154 (1952).
[17]J. D. Roberts and R. H. Mazur, *J. Amer. Chem. Soc.*, **73**, 3542 (1951).

Comments

Only recently has the proper interpretation of the solvolytic behavior of many systems with neighboring groups become clear.[18,19] More nucleophilic solvents, such as aqueous alcohols or acetic acid, can attack substrate in S_N2-type processes. Such solvent participation often competes very successfully against neighboring group participation with the result that only small rate enhancements (or even rate depressions) are observed in comparison with model compounds. Winstein's choice of acetolysis as "standard" for the comparison of tosylate rates was unfortunate, since the true magnitude of anchimeric assistance was often masked. Thus, the $C_6H_5CH_2CH_2OTs/CH_3CH_2OTs$ rate ratio varies 9000-fold with change of solvent, from 0.20 in 50% ethanol, through 0.35 in acetic acid and 2.7 in formic acid, to 1770 in trifluoroacetic acid.[19] Only the last value reveals the true participating ability of the β-phenyl group. Appreciation of the importance of solvent assistance led to the resolution of the phenonium ion problem.[20] It is important to realize that anchimeric assistance need not be of large magnitude to be significant. For example, a k_Δ/k_s ratio of only 5 implies that 80% of the reaction proceeds through the anchimerically assisted pathway.

Although it is now possible to determine the magnitude of solvent assistance quantitatively,[18,19] the simplest approach is, whenever possible, to compare data in limiting solvents such as trifluoroacetic acid[21] or hexafluoroisopropanol.[22] This is very important for primary and unhindered or modestly hindered secondary substrates, where solvent participation can be important. Tertiary systems not prone to S_N2 or E2 attack by solvent do not present a problem in this regard.

The less stable the carbocation, the more important participation effects are likely to be. In the absence of solvent participation, anchimeric

[18]See Chapter 11.2 and Refs. 10 and 11 therein. T. W. Bentley and P. v. R. Schleyer, *J. Amer. Chem. Soc.*, **98**, 7658 (1976); T. W. Bentley, F. L. Schadt, and P. v. R. Schleyer, *ibid.*, **98**, 7667 (1976).

[19]F. L. Schadt and P. v. R. Schleyer, *J. Amer. Chem. Soc.*, **95**, 7861 (1973); A. Diaz, I. Lazdins, and S. Winstein, *ibid.*, **90**, 6546 (1968).

[20]Compare H. C. Brown, "The Transition State," *Chem. Soc.* (*London*) *Spec. Publ.* **16**, 149 (1962), and *J. Amer. Chem. Soc.*, **87**, 2137 (1965) with P. v. R. Schleyer and C. J. Lancelot, *ibid.*, **91**, 4297 (1969) and H. C. Brown, C. J. Kim, C. J. Lancelot, and P. v. R. Schleyer, *ibid.*, **92**, 5244 (1970). Review: C. J. Lancelot, D. J. Cram, and P. v. R. Schleyer, *Carbonium Ions*, Vol. III, G. Olah and P. v. R. Schleyer, Eds., Wiley-Interscience, New York, 1972, Chapter 27, p. 1347ff.

[21]A. C. Cope, J. M. Grisar, and P. E. Peterson, *J. Amer. Chem. Soc.*, **81**, 1640 (1959); **82**, 4299 (1960); P. E. Peterson, *J. Amer. Chem. Soc.*, **82**, 5834 (1960).

[22]F. L. Schadt, P. v. R. Schleyer, and T. W. Bentley, *Tetrahedron Lett.*, 2335 (1974).

assistance is most likely in primary substrates, least likely in tertiaries. Steric assistance should exhibit just the opposite order, and Brown has convincingly demonstrated that most (but not all) rate accelerations in tertiary substrates have a steric, rather than a participative origin. Extrapolation of this interpretation to the behavior of secondary systems, however, is open to doubt.

It should not be assumed that σ-groups are always inherently poorer neighboring groups than π- and n-donors. A clearcut distinction among these categories is not always evident. Neighboring group ability should mirror basicity, and it is instructive to compare the inherent basicities of compounds in the gas phase, where solvation effects are not present. Thus, the proton affinity (PA) of HF (112 kcal/mole),[23] a n-donor, is actually less than that of CH_4 (126 kcal/mole),[24] a σ-donor. The strained σ-bond of cyclopropane is remarkably basic. The proton affinity of cyclopropane (179 kcal/mole, to give protonated cyclopropane)[25] actually exceeds that of the π-donor, ethylene (PA = 159 kcal/mole),[26] and even that of water (PA = 165 kcal/mole)![27]

Professor Brown and I agree that many unambiguous cases of σ-*participation* are known in unsymmetrical systems where rearrangement to a more stable ion takes place during reaction. Thus, solvolysis of 1-methylcyclobutylcarbinyl tosylate is enormously accelerated,[28] but the driving force is provided by the relief of cyclobutane strain and the conversion of a primary system to a tertiary ion. No bridged *intermediate* need intervene in such instances.

RR (HOAc, 25°C):	1.0	24,600

[23]M. S. Foster and J. L. Beauchamp, *Inorg. Chem.*, **14**, 1229 (1975).

[24]W. A. Chupka and J. Berkowitz, *J. Chem. Phys.*, **54**, 4256 (1971); M. A. Haney and J. L. Franklin, *Trans. Faraday Soc.*, **65**, 1794 (1969).

[25]S.-L. Chong and J. L. Franklin, *J. Amer. Chem. Soc.*, **94**, 6347 (1972); D. J. McAdoo, F. W. McLafferty, and P. F. Bente III, *ibid.*, **94**, 2027 (1972).

[26]J. L. Franklin, J. D. Dillard, H. M. Rosenstock, J. T. Herron, K. Draxl, and F. H. Field, *Nat. Stand. Ref. Data Ser. Nat. Bur. Stand.*, No. 26 (1969).

[27]M. A. Haney and J. L. Franklin, *J. Phys. Chem.*, **73**, 4328 (1969).

[28]C. Woodworth, Ph.D. Thesis, Princeton University, 1969. See P. v. R. Schleyer and E. Wiskott, *Tetrahedron Lett.*, 2845 (1967).

4

Carbon-Bridged Cations

4.1. Introduction

In the last chapter, the evidence for π-, aryl-, and σ-participation in representative systems was examined. Unambiguous evidence for both π- and aryl participation was found, but no unambiguous evidence for σ-participation was uncovered in symmetrical systems not undergoing rearrangement to more stable structures. However, detailed consideration of 2-norbornyl and cyclopropylcarbinyl, considered at one time to be the best examples for such σ-participation, was deferred until later (Chapters 5 and 6).

We now turn our attention to the nature of the intermediates produced in the solvolysis of these derivatives. Ordinarily, we would concern ourselves only with those systems where the rate data reveal significant participation. According to the Hammond postulate, the transition state in a solvolytic process should resemble closely the first intermediate.[1] However, it has been proposed that "there are good reasons to expect carbon bridging to lag behind C–X ionization at the transition state."[2] Consequently, it is necessary to examine the intermediate produced for carbon bridging even in cases where no evidence for carbon participation in the transition state has been detected.

4.2. What Is a Nonclassical Ion?

Surprising as it may seem, in view of the vast literature on the subject, there appears to have been no serious attempt by workers in the field to

[1]G. S. Hammond, *J. Amer. Chem. Soc.*, **77**, 334 (1955).
[2]S. Winstein, *J. Amer. Chem. Soc.*, **87**, 381 (1965).

to arrive at a generally acceptable definition of the term "nonclassical ion." Consequently, it appears appropriate to turn back to the way the term was introduced and used to arrive at an unambiguous definition.

$$\text{(1)}$$

1   **2**

The term "nonclassical" was apparently first used by Roberts some twenty-five years ago in referring to his proposed tricyclobutonium structure **2** for the cyclopropylcarbinyl cation **1**[3] (1). The proposed structure **2** was clearly different from the "classical" structure **1** for the intermediate cation.

The second time the term was apparently used by Winstein, who refers to the "nonclassical structures" of the norbornyl cation (**4**) and other cations in contrast to their classical structures (**3**)[4] (2).

$$\text{(2)}$$

3   **4**

The third such reference is apparently due again to Roberts: "Recent interest in the structures of carbonium ions has led to speculation as to whether the ethyl cation is most appropriately formulated as a simple solvated electron-deficient entity (**5**), a "nonclassical" bridged ethylene protonium ion (**6**), or possibly as an equilibrium mixture of the two ions"[5] (3).

$$\text{(3)}$$

5   **6**

Although in the past the term has also been applied to n-bridged species, such as the bromonium ion[4] and to π-bridged species, such as the phenonium ion,[4] it has been urged by both Bartlett[6] and Sargent[7] that the

[3]J. D. Roberts and R. H. Mazur, *J. Amer. Chem. Soc.*, **73**, 3542 (1951).
[4]S. Winstein and D. Trifan, *J. Amer. Chem. Soc.*, **74**, 1147, 1154 (1952).
[5]J. D. Roberts and J. A. Yancy, *J. Amer. Chem. Soc.*, **74**, 5943 (1952).
[6]P. D. Bartlett, *Nonclassical Ions*, Benjamin, New York, 1965.
[7]G. D. Sargent, *Carbonium Ions*, Vol. III, G. A. Olah and P. v. R. Schleyer, Eds., Wiley–Interscience, New York, 1972, Chapter 24.

term "nonclassical" be restricted to delocalizations involving σ-electrons. After all, there is nothing nonclassical about the bromonium ion (**7**). It can be nicely represented by a single Lewis structure.

$$
\begin{array}{c}
\overset{+}{Br} \\
\diagup \ \diagdown \\
H_2C \!-\! CH_2 \\
\mathbf{7}
\end{array}
$$

Similarly, there would appear to be nothing exceptional requiring the nonclassical designation about a protonated aromatic (**8**)[8] or the heptamethylbenzene cation (**9**)[9] (4).

(4)

Both Sargent and Ingold conceived of the bridged ions as involving a strong interaction between carbon–carbon (or carbon–hydrogen) σ electrons and the cationic center.[7,10] Ingold introduced the term, "synartetic ion," for such species, which he described as possessing a "split single bond *fastening together* the locations of a split ionic charge."[10]

Actually, there are major advantages in simplicity in limiting the term "nonclassical carbonium ion," to cations containing carbon or hydrogen bridging atoms, such as **2**, **4**, and **6**. Accordingly, in a collective effort, Professor Paul von R. Schleyer and the author have developed the following definitions restricted to such cations.

1. A carbonium ion (carbocation) is a positively charged species in which a significant portion of the positive charge resides on one or more carbon atoms termed the "carbonium" carbon or carbons.

2. A classical carbonium ion is a positively charged species which can be adequately represented by a single Lewis structure involving only two electron–two center bonds. (Typical examples are **1**, **3**, and **5**.) Traditionally, π-conjugated cations, such as allyl and cyclopropenyl, are included in this category.

[8]H. C. Brown, H. W. Pearsall, L. P. Eddy, W. J. Wallace, M. Grayson, and K. L. Nelson, *Ind. Eng. Chem.*, **45**, 1462 (1953).

[9]W. v. E. Doering, M. Saunders, H. G. Boyton, H. W. Earhart, E. F. Wadley, W. R. Edwards, and G. Laber, *Tetrahedron*, **4**, 178 (1958).

[10]F. Brown, E. D. Hughes, C. K. Ingold, and J. F. Smith, *Nature*, **168**, 65 (1951).

3. A nonclassical carbonium ion is a positively charged species which cannot be represented adequately by a single Lewis structure. Such a cation contains one or more carbon or hydrogen bridges joining the two electron-deficient centers. The bridging atoms have coordination numbers higher than usual, typically five or more for carbon and two or more for hydrogen. Such ions contain two electron–three (or multiple) center bonds including a carbon or hydrogen bridge. (Typical examples are **2**, **4**, and **6**.)

Some authors have termed homoconjugated carbocations "nonclassical." However, they have neither bridges, nor two electron–three (or multiple) center bonds. Consequently, I believe that it is not desirable to apply the nonclassical designation to such homoconjugated species. Indeed, they are not pertinent to the question explored in this book—how important are σ-bridged cations in the solvolytic behavior of organic derivatives and in the representative reactions of carbonium ions in the usual organic media.

4.3. σ-Bridging vs. Hyperconjugation

A definition is satisfactory when it divides selected species into a category with well recognized common features. However, chemistry is essentially continuous. Almost any definition encounters borderline cases which are difficult to classify.

Thus, chemical bonds can be clearly separated into two major classes: ionic and covalent. But what do we do with covalent bonds with partial ionic character or ionic bonds with partial covalent character?[11]

A similar difficulty is encountered in attempting to differentiate between σ-bridging and hyperconjugation. Thus Bartlett[6] has proposed the definition: "an ion is nonclassical if its ground state has delocalized bonding σ electrons." But essentially all aliphatic, alicyclic, and bicyclic cations (with the exception of the methyl cation, CH_3^+) are stabilized by hyperconjugation. Hyperconjugation involves delocalized σ electrons. Would it be helpful to call Roberts' "classical" structure for the ethyl cation (**5**) "nonclassical"? Or the relatively stable *tert*-butyl cation?

Perhaps it would clarify things to review briefly the theoretical situation as it existed just prior to the introduction of the nonclassical concept.

In the 1940s, electronic effects, such as inductive, inductomeric, field, mesomeric, and hyperconjugative, were generally recognized and utilized to interpret structural effects in organic chemistry. The book by my former colleague at Wayne University, A. E. Remick,[12] published in 1943 provides

[11]L. Pauling, *The Nature of the Chemical Bond*, 3rd ed., Cornell University Press, Ithaca, N.Y., 1960.
[12]A. E. Remick, *Electronic Interpretations of Organic Chemistry*, Wiley, New York, 1943.

a convenient summary of the *status quo* prior to the nonclassical ion proposal.

The *tert*-butyl cation is much more stable than the methyl cation. We account for this today in terms of the hyperconjugative contributions which stabilize the *tert*-butyl cation and delocalize charge from the carbonium carbon into the methyl groups (5).

$$
\underset{\underset{\text{CH}_3}{|}}{\overset{\overset{\text{CH}_3}{|}}{\text{H}_3\text{C}-\text{C}^+}} \longleftrightarrow \underset{\underset{\text{H}\quad\text{CH}_3}{|\quad|}}{\overset{\overset{\text{H}\quad\text{CH}_3}{|\quad|}}{\text{H}^+\text{C}=\text{C}}} \tag{5}
$$

(9 structures)

Does such delocalization of charge into a saturated alkyl group correspond to the new concept proposed in the 1950s and attributed to nonclassical carbonium ions?

Let us see what Remick had to say on the subject in 1943 (p. 394).

> "On the basis of these arguments, one might wonder how a secondary or tertiary alkyl halide can yield a carbonium ion which is sufficiently stable to permit of a S_N1 reaction. The explanation is readily found in the concept of hyperconjugation. Thus no stabilization by resonance is possible in the carbocation produced from methyl bromide, whereas in ethyl bromide, it becomes conceivable,
>
> $$
> \underset{\underset{\text{H}\quad\text{H}}{|\quad|}}{\overset{\overset{\text{H}\quad\text{H}}{\scriptstyle\curvearrowright\,|}}{\text{H}-\text{C}-\text{C}^+}} \qquad \text{Br}^-
> $$
>
> and in secondary and tertiary halides, it becomes still more probable owing to the increasing number of contributing structures."

It was against this background that σ-bridging was introduced as a new phenomenon which provided an additional stabilizing factor for carbonium ions.[3-5]

Consequently, to take a specific example, a classical carbonium ion, such as ethyl (5) involves a certain amount of hyperconjugative stabilization involving the β-hydrogen. Such stabilization will cause the β-hydrogen to move toward the carbonium ion. We could describe the electronic stabilization as involving weak σ-bridging. At what stage should we consider that such σ-bridging has become sufficiently important so as to deserve formulation as a σ-bridged nonclassical cation?

Ingold discussed such cases.[13] He pointed out that in a Wagner–

[13]C. K. Ingold, *Structure and Mechanism in Organic Chemistry*, 2nd ed., Cornell University Press, Ithaca, N.Y., 1969, pp. 766–787.

Meerwein equivalent pair (6) the two cations (**10, 10′**) pass over a low barrier which separates the two minima in the energy coordinate. The top of the

$$
\underset{\textbf{10}}{-\overset{R}{\underset{|}{C}}-\overset{+}{\underset{|}{C}}-} \;\rightleftharpoons\; \underset{\textbf{10′}}{-\overset{+}{\underset{|}{C}}-\overset{R}{\underset{|}{C}}-}
\tag{6}
$$

barrier then corresponds to a symmetrical species (**11**) with the migrating group, R, half-way between the two carbon termini.

$$
\underset{\textbf{11}}{-\underset{|}{C}\overset{\overset{R}{\diagup+\diagdown}}{-\!\!-}\underset{|}{C}-}
$$

The nonclassical (synartetic) proposal is that the barrier disappears, so that resonance takes place between the canonical structures equivalent to **10** and **10′**, resulting in a stabilized resonance hybrid equivalent to **11**.

The quantum mechanical perturbation treatment of F. K. Fong is more elegant[14,15] although the net result appears to be the same. It involves application of his new rate theory to the treatment of Wagner–Meerwein rearrangements in carbocations with special focus on the low-barrier limit of double-minimum relaxation. In this approach, hypothetical structures, without any electronic (hyperconjugative) interactions arising from the presence of the positive charge, are taken to be the zeroth order states. Inclusion of these interactions gives rise to a smooth transition of low-barrier equilibrating pairs (**10 ⇌ 10′**) that becomes a single symmetrical species (**11**) in the zero-barrier limit. The sole clear division between ions which undergo hyperconjugative stabilization in rapidly equilibrating pairs (**10 ⇌ 10′**) and those which are so stabilized as to give a symmetrical bridged species (**11**) is between cations which exhibit double (or multiple) minima in their energy coordinate and those which exhibit single minima in that coordinate.

The situation is less clear for unsymmetrical Wagner–Meerwein cationic pairs.[13–15] Consequently, in this study we shall emphasize symmetrical systems in so far as that is practical.

Let us consider some other problems. Methane accepts a proton in the gas phase (under mass-spec. conditions) to form a relatively stable species, CH_5^{+}.[16] (Similar protonated species, such as $C_2H_7^{+}$, H_3^{+}, and NeH^{+}, are also known.[17]) Is CH_5^{+} a nonclassical carbonium ion? Let us consider its structure to see if it answers our criteria for such a species.

[14]F. K. Fong, *J. Amer. Chem. Soc.*, **96**, 7638 (1974).
[15]F. K. Fong, *Theory of Molecular Relaxation*, Wiley–Interscience, New York, 1975, Chapter 8.
[16]F. H. Field and M. S. B. Munson, *J. Amer. Chem. Soc.*, **87**, 3289 (1965).
[17]F. H. Field, *Mass Spectroscopy*, Vol. V, A. Maccoll, Ed., Butterworths, London, 1972.

One possibility is that the proton is merely associated with the methane molecule as a whole, wandering about the CH_4 unit, with no significant minima. There would then be no specific σ-bridge involving a carbonium ion center.

$$
\left[
\begin{array}{c}
\text{H} \\
\text{H}\cdots\overset{\vdots}{\underset{\vdots}{\text{C}}}\cdots\text{H} \\
\text{H}\quad\text{H}
\end{array}
\right]^{+}
\qquad
\begin{array}{c}
\text{H} \\
\text{H}\diagdown\overset{|}{\underset{\uparrow}{\text{C}}}\diagup\text{H} \\
\text{H}\!-\!\text{H}
\end{array}
\equiv
\begin{array}{c}
\text{H} \\
\text{H}\diagdown\overset{|}{\text{C}}\diagup\text{H} \\
\text{H}\cdots\overset{+}{\cdots}\text{H}
\end{array}
\qquad (7)
$$
$$
\quad\;\;\mathbf{12}\qquad\qquad\qquad\mathbf{13}
$$

Another possibility is that the species formed contains a carbon atom bonded to five individual hydrogen atoms (**12**) (7). Should we consider such a species to be a nonclassical carbonium ion? The carbon atom does have five nearest neighbors. Thus the carbon atom exhibits the higher coordination number of a bridging carbon. On the other hand, the structure does not involve a σ-bridge between two electron-deficient centers.

Alternatively, it has been proposed on the basis of calculations that the structure of CH_5^+ is better represented as a methyl carbonium ion coordinated with a hydrogen molecule (**13**).[18] This species does involve a σ-bridge between two electron-deficient centers. It clearly contains a two electron–three center bond similar to that present in H_3^+. Consequently, both CH_5^+ and H_3^+ are nonclassical ions, but strictly speaking, are not nonclassical carbonium ions in accordance with the proposed definition. However, possibly we should stretch a point and admit CH_5^+ as a prototype nonclassical carbonium ion.

Let us consider the next higher member, $C_2H_7^+$.[19] Is this a nonclassical ion? The evidence is that CH_5^+ is a species of considerable stability[18]:

$$
\text{H}_3\overset{+}{\text{C}}\rightarrow\overset{\text{H}}{\underset{\text{H}}{|}}
\;\rightleftharpoons\;
\text{H}_3\text{C}^+ + \text{H}_2
\qquad \Delta H = \sim 40\,\text{kcal mol}^{-1}
\qquad (8)
$$

On the other hand, $C_2H_7^+$ is a weak complex, readily dissociated into its components[19]:

$$
\text{CH}_3\text{CH}_2^+\rightarrow\overset{\text{H}}{\underset{\text{H}}{|}}
\;\rightleftharpoons\;
\text{CH}_3\text{CH}_2^+ + \text{H}_2
\qquad \Delta H \approx 0\,\text{kcal mol}^{-1}
\qquad (9)
$$

It does not appear desirable to include such weak complexes in the category of nonclassical carbonium ions, a class of cations which were considered to be highly stabilized relative to the corresponding classical structures.

---

[18]V. Dyczmons and W. Kutzelnigg, *Theor. Chim. Acta*, **33**, 239 (1974).
[19]S. Chong and J. L. Franklin, *J. Amer. Chem. Soc.*, **94**, 6347 (1972).

Recent calculations indicate that in the gas phase the stable form of the *n*-propyl cation (**14**) may be the symmetrically bridged species[20] (**15**). Clearly this would be a nonclassical carbonium ion.

$$CH_3CH_2CH_2^+ \qquad \underset{\textstyle H_2C\text{---}CH_2}{\overset{\textstyle \overset{H_3}{C}}{}}$$

CH₃CH₂CH₂⁺ $\begin{array}{c} H_3 \\ C \\ {}^{+} \\ H_2C\!=\!=\!CH_2 \end{array}$

14   **15**

The corner-protonated cyclopropane species (**15**), or the edge-protonated cyclopropane species (**16**), also would constitute authentic examples of nonclassical ions.[21]

$\begin{array}{c} H_2 \\ C \\ / \quad \backslash \\ H_2C\text{----}CH_2 \\ {}^{+} \\ H \end{array}$

16

However the original proposals were for nonclassical structures for cations produced in solution.[3-5] The precise structure of cations in the gas phase is an interesting question, but a different one.[22] There is no satisfactory evidence at present that nonclassical cations, such as protonated hydrogen, H_3^+, or protonated methane, CH_5^+, can exist in the media used to form carbonium ions and to observe their reactions, in spite of some highly interesting and imaginative discussions.[23,24] Similarly, the *n*-propyl cation appears to be generated in solvolytic media primarily in its unbridged form.[25,26] In the present discussion, we will restrict our examination primarily to the structure of cations produced under solvolytic or stable ion conditions.

4.4. Static Classical, Equilibrating Classical, or σ-Bridged Cations

In Figure 4.1 are shown several representative carbonium ions for which σ-bridged nonclassical structures have been considered. In such cases

[20]L. Radom, J. A. Pople, V. Buss, and P. v. R. Schleyer, *J. Amer. Chem. Soc.*, **94**, 311 (1972); P. C. Hariharan, L. Radom, J. A. Pople, and P. v. R. Schleyer, *ibid.*, **96**, 599 (1974).

[21]M. Saunders, P. Vogel, E. L. Hagen, and J. Rosenfeld, *Accounts Chem. Res.*, **6**, 53 (1973).

[22]H. C. Brown, "The Transition State," *Chem. Soc. (London) Spec. Publ.*, **16**, 140–158, 174–178 (1962); W. L. Jorgensen, *J. Amer. Chem. Soc.*, **99**, 280 (1977).

[23]G. A. Olah, J. Shen, and R. H. Schlosberg, *J. Amer. Chem. Soc.*, **95**, 4957 (1973).

[24]G. A. Olah, G. Klopman, and R. H. Schlosberg, *J. Amer. Chem. Soc.*, **91**, 3261 (1969).

[25]J. D. Roberts and M. Halmann, *J. Amer. Chem. Soc.*, **75**, 5759 (1953).

[26]O. A. Reutov and T. N. Shatkina, *Tetrahedron*, **18**, 327 (1962).

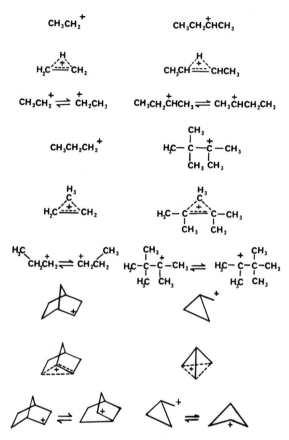

Figure 4.1. Representative carbonium ions for which σ-bridged nonclassical structures have been considered.

we have three typical possibilities to consider. First, under the reaction conditions the carbonium ion is formed and reacts in the static classical form, shown at the top level of each group. Second, the classical carbonium ion is formed and undergoes rapid equilibration, as shown in the lowest level of each group. Such equilibration may be slow, comparable with, or faster than the rate of capture by the nucleophile. Third, the σ-bridged ion may be formed and is the species that reacts with the nucleophile.

In these systems, the σ-bridged nonclassical structure possesses a plane of symmetry not present in the classical structure. Consequently, by introducing a tag in an appropriate position and then examining the reaction product, it is possible to test each system for the formation or nonformation of such a symmetrical species.

Indeed, investigation soon revealed that the ethyl,[5] *n*-propyl,[25,26] *sec*-butyl,[27] and 2,3,3-trimethyl-2-butyl[28] cations can be generated and transformed into product without full equilibration of the tag.

One example of a study of this kind may be described in some detail. The solvolysis of the tagged 2,3,3-trimethyl-2-butyl 3,5-dinitrobenzoate (**17**) in methanol at 100°C exhibits typical S_N1 behavior. The product is both olefin and the methyl ether (**19**)[29] (10). The methyl ether exhibits no

$$
\underset{\textbf{17}}{\overset{\displaystyle H_3C\ \ CD_3}{\underset{\displaystyle H_3C\ \ O_2CAr}{H_3C-\overset{|}{\underset{|}{C}}-\overset{|}{\underset{|}{C}}-CD_3}}}
\xrightarrow[100°C]{CH_3OH}
\underset{\textbf{18}}{\overset{\displaystyle H_3C\ \ CD_3}{\underset{\displaystyle H_3C}{H_3C-\overset{|}{\underset{|}{C}}-\overset{|}{\underset{+}{C}}-CD_3}}}
\xrightarrow{CH_3OH}
\underset{\textbf{19}}{\overset{\displaystyle H_3C\ \ CD_3}{\underset{\displaystyle H_3C\ \ OCH_3}{H_3C-\overset{|}{\underset{|}{C}}-\overset{|}{\underset{|}{C}}-CD_3}}}
\quad (10)
$$

detectable scrambling of the tag. Consequently, the intermediate must be neither the bridged species (**20**), nor a pair of classical ions that equilibrates (**21**) competively with the rate of capture (11).

$$
\underset{\textbf{20}}{\underset{\displaystyle H_3C\ \ \ \ CD_3}{H_3C-C\overset{\overset{\displaystyle H_3}{\displaystyle C}}{\cdots\overset{+}{\cdots}}C-CD_3}}
\qquad
\underset{\textbf{21}}{\overset{\displaystyle H_3C\ \ CD_3}{\underset{\displaystyle H_3C}{H_3C-\overset{|}{\underset{|}{C}}-\overset{|}{\underset{+}{C}}-CD_3}}}
\rightleftharpoons
\overset{\displaystyle H_3C\ \ CH_3}{\underset{\displaystyle CD_3}{H_3C-\overset{|}{\underset{+}{C}}-\overset{|}{\underset{|}{C}}-CD_3}}
\quad (11)
$$

But what decision can be reached for ions such as cyclopropylcarbinyl[3] and 2-norbornyl,[4] where appropriate tags are fully scrambled? How can we decide between the proposed nonclassical formulations, (**2**) and (**4**), and the corresponding formulation as rapidly equilibrating classical cations (Figure 4.1)?

The cyclopropylcarbinyl question has proved a relatively easy one to answer (Chapter 5). However, the 2-norbornyl question (Chapter 6) has proved to be much more difficult. Indeed, much of the remainder of the book will be devoted to various experimental attacks on that question.

4.5. Stereochemistry and σ-Bridged Cations

In cases where the symmetry test, discussed in the previous section, cannot be used, recourse has been had to stereochemistry to support proposals for σ-bridging. Thus the great majority of secondary and tertiary derivatives undergo solvolysis with inversion or predominant racemization

[27] J. D. Roberts, W. Bennett, R. E. McMahon, and E. W. Holroyd, Jr., *J. Amer. Chem. Soc.*, **74**, 4283 (1952).

[28] J. D. Roberts and J. A. Yancey, *J. Amer. Chem. Soc.*, **77**, 5558 (1955).

[29] H. C. Brown and C. J. Kim, *J. Amer. Chem. Soc.*, **90**, 2082 (1968).

of the reaction center.[30] Observation of a solvolysis with preferential retention has led to proposals for σ-participation and σ-bridged cations.

For example, 7-norbornyl esters (**22**) are remarkably inert. They undergo solvolysis in acetic acid (12) only very slowly under remarkably drastic conditions to give predominantly 7-norbornyl acetate (**23**) with a relatively minor amount of *exo*-2-bicyclo[3.2.0]heptyl acetate (**24**).[31]

$$(12)$$

By means of 2,3-dideuterium labeling, it was shown that this solvolysis proceeds with predominant retention.[32,33] Consequently, it was proposed that the incipient ion is bridged[32-34] (13).

$$(13)$$

Unfortunately for this proposal, there is no basis to postulate any stabilization of this intermediate. Thus the rate of acetolysis of the parent compound is predicted correctly by the Foote–Schleyer correlation.[35]

One of the problems may be the naive extrapolation of the stereochemical results realized with the relatively simple systems that have been explored to the much more complex systems, such as these bicyclics.

In systems such as 2-octyl tosylate, conversion to an ion pair leaves an intermediate where solvent is readily captured from the backside to produce a product with inversion.[36,37] However, in a system which is sterically

[30]W. J. le Noble, *Highlights of Organic Chemistry*, M. Dekker, New York, 1974.

[31]S. Winstein, F. Gadient, E. T. Stafford, and P. E. Klinedinst, Jr., *J. Amer. Chem. Soc.*, **80**, 5895 (1958).

[32]P. G. Gassman and J. M. Hornback, *J. Amer. Chem. Soc.*, **89**, 2487 (1967).

[33]F. B. Miles, *J. Amer. Chem. Soc.*, **89**, 2488 (1967); *ibid.*, **90**, 1265 (1968).

[34]M. J. S. Dewar and W. W. Schoeller, *Tetrahedron*, **27**, 4401 (1973).

[35]P. v. R. Schleyer, *J. Amer. Chem. Soc.*, **86**, 1856 (1964).

[36]R. A. Sneen, *Accounts Chem. Res.*, **6**, 46 (1973).

[37]D. J. Raber, J. M. Harris, and P. v. R. Schleyer, *Ions and Ion Pairs in Organic Reactions*, Vol. II, M. Szwarc, Ed., Wiley, New York, 1974.

crowded, approach of solvent from the backside to the ion pair might become relatively slow (**26**). Then solvent capture from the front side, possibly solvent molecules associated with the anion, could become important.[38] This would result in retention. For example, Goering and Chang have found that the solvolysis of active 2-phenyl-2-butyl *p*-nitrobenzoate (**25**) in aqueous acetone proceeds to give alcohol (**27**) with as much as 65% of the retained optically active structure[39] (14).

$$
\text{H}_3\text{CCHCH}_2\text{CH}_3 \longrightarrow \text{H}_3\overset{+}{\text{C}}\text{CHCH}_2\text{CH}_3 \xrightarrow{\text{H}_2\text{O}} \text{H}_3\text{CCHCH}_2\text{CH}_3 \qquad (14)
$$

OPNB	⁻OPNB	OH
25	**26**	**27**

Substitution processes involving inversion are extraordinarily difficult in rigid bicyclic structures such as norbornane. It is entirely possible that the 7-norbornyl brosylate goes to a tight ion pair without σ-bridging. This could now undergo solvent capture from the solvent molecules associated with the anion portion of the ion pair,[37] as well as undergo Wagner–Meerwein shifts. In this way the results could be accounted for without postulating σ-bridges (13) which are not compatible with the exceptional inertness of the system.

Similarly, *exo*-2-phenylnorbornyl chloride (**28**) undergoes solvolysis (15) to 2-phenyl-*exo*-norbornanol (**30**) with complete retention.[40] Yet, the highly stable 2-phenyl-2-norbornyl cation (**29**) is very stable and cannot be σ-bridged.[41]

$$ \text{Cl} \longrightarrow \xrightarrow{\text{H}_2\text{O}} \text{OH} \qquad (15) $$

28   **29**   **30**

Consequently, stereochemical retention should be used with caution as a basis for postulating σ-bridged cations.

[38]H. C. Brown, K. L. Morgan, and F. J. Chloupek, *J. Amer. Chem. Soc.*, **87**, 2137 (1965).

[39]H. L. Goering and S. Chang, *Tetrahedron Lett.*, 3607 (1965). S. Chang, Ph.D. Thesis, University of Wisconsin, 1966.

[40]H. C. Brown, F. J. Chloupek, and M.-H. Rei, *J. Amer. Chem. Soc.*, **86**, 1246 (1964).

[41]D. G. Farnum and G. Mehta, *J. Amer. Chem. Soc.*, **91**, 3256 (1969).

4.6. π- and Aryl-Bridged Cations

Even though π- and aryl-bridged cations are not closely associated with the question of σ-bridged (nonclassical) cations, it might be helpful to consider here some of the fascinating phenomena involving those cations in order to clarify some of the unfortunate misunderstandings that have plagued this area of research.

The presence of a double bond in norbornene *anti* to the 7-substituent enormously enhances the rate of solvolysis[42,43] (16). Clearly the solvolyses

$$\text{(16)}$$

RR (25°C):   1.00 10^{11} 10^{14}

are proceeding with participation of the double bond (17). The question is whether the intermediate produced is a pair of rapidly equilibrating cyclo-

$$\text{(17)}$$

propylcarbinyl cations (**34**), with **35** as a stabilized transition state, or is **35** a species much more stable than **34**, so stable that we need no longer consider these structures.

The situation is even more complicated for **33** (18).

$$\text{(18)}$$

---

[42]S. Winstein, M. Shatavsky, C. Norton, and R. B. Woodward, *J. Amer. Chem. Soc.*, **77**, 4183 (1955).
[43]S. Winstein and C. Ordronneau, *J. Amer. Chem. Soc.*, **82**, 2084 (1960).

We undertook to test **37** or **38** relative to **36**, by trapping the ion with sodium borohydride.[44] We argued that the *endo* transannular bonding proposed for **37** and **38** should resist capture of the borohydride by the

$$\text{(structures)} \xrightarrow[\textbf{40}]{\text{NaBH}_4} \quad + \quad \tag{19}$$

83%   12%

cation from the *endo* direction. However, we obtained an 83% yield of the tricyclic hydrocarbon **40** predicted for the intermediate **36** (19). Solvolysis of **32** in the presence of sodium borohydride also gave the tricyclic hydrocarbon **41**, but in a lower yield (20).

$$\text{(structures)} \xrightarrow[\textbf{41}]{\text{NaBH}_4} \quad + \quad + \quad \tag{20}$$

15%   70%   6%

Solvolysis of **32** and **33** in methanol yields the *anti*-7-methyl ethers (**42** and **44**, respectively). However, in strongly basic methanol, the tricyclic derivatives (**43** and **45**, respectively), are obtained[45,46] (21).

42 $\xleftarrow{\text{CH}_3\text{OH}}$ **36 or 37** $\xrightarrow{\text{NaOCH}_3}$ **43**

(21)

44 $\xleftarrow{\text{CH}_3\text{OH}}$ **34 or 35** $\xrightarrow{\text{NaOCH}_3}$ **45**

A possible explanation for these results is provided by the results of J. J. Tufariello.[47] He achieved the synthesis of the tricyclic alcohol (**46**). This was

[44]H. C. Brown and H. M. Bell, *J. Amer. Chem. Soc.*, **85**, 2324 (1963).

[45]H. Tanida, T. Tsuji, and T. Irie, *J. Amer. Chem. Soc.*, **88**, 864 (1966).

[46]A. Diaz, M. Brookhart, and S. Winstein, *J. Amer. Chem. Soc.*, **88**, 3133 (1966).

[47]J. J. Tufariello, T. F. Mich, and R. J. Lorence, *Chem. Commun.*, 1202 (1967); J. J. Tufariello and R. J. Lorence, *J. Amer. Chem. Soc.*, **91**, 1546 (1969).

perfectly stable in buffered solution, but in even weak acid, it underwent very rapid transformation to *anti*-7-norbornenol (**47**) (22). The enormous

$$\qquad\qquad (22)$$

reactivity of the tricyclic structure **46** is indicated by the fact that its derivatives undergo solvolysis at a rate 10^{16} greater than the related *endo*-norbornyl compounds.

Presumably, the first product of the reaction of methanol or water with the cation (**34**) must be the protonated species (**48**) (23). Rupture of the

$$\qquad\qquad (23)$$

highly reactive cyclopropylcarbinyl bond must be extraordinarily fast, faster even than the transfer of the proton. In other words, the equilibrium involving the protonated species **48** must favor the free cation, which then ends up as the more stable *anti*-7-derivative. Nucleophiles, such as BH_4^-, CN^-, and $^-OCH_3$, which give more stable unprotonated derivatives, provide the predicted tricyclic derivatives (**40, 41, 43, 45**).

In an attempt to decide the question of whether the intermediate is the rapidly equilibrating set of unsymmetrical tricyclic cations (**34**), or the symmetrical bridged species **35**, Gassman introduced one and two methyl groups on the double bond of **32**. The relative rates (24) reveal that the effect

$$\qquad\qquad (24)$$

| RR (25°C): | 1.00 | 13 | 150 |

is a symmetrical one in the transition state.[48] The problem is whether we can extrapolate the symmetrical effects in the transition state to postulate a symmetrical intermediate.

[48]P. G. Gassman and D. S. Patton, *J. Amer. Chem. Soc.*, **91**, 2160 (1969).

In fact, Bartlett and Sargent encountered difficulties with a similar extrapolation.[49] The solvolysis of 2-(Δ^3-cyclopentenyl)-ethyl nosylates (**49**) proceeds with participation to give norbornyl cations (**50 or 51**) as intermediates (25). Since the rates increased in a regular manner from 1.00

$$\text{(25)}$$

49   **50**   **51**

(R, R' = H), to 7 (R = Me, R' = H), to 38 (R, R' = Me), the substituent effect is a symmetrical one. It was then postulated that a reaction starting from a symmetrical ground state (**49**) and passing through a symmetrical transition state must produce a symmetrical initial intermediate. Alas! It was only shortly afterward that Goering and Humski successfully trapped the optically active 1,2-dimethyl-2-norbornyl cation[50] (**51**) (R, R' = Me). Consequently, it cannot be the symmetrically bridged species predicted.[49]

The 7-benzonorbornenyl system provides a convincing case for aryl-participation[51] (26). Substituents in the 2',3'-positions reveal a symmetrical

$$\text{(26)}$$

RR (77.6°C):   55   1.00   0.00014

effect upon the rate (3'-Me, 6; 2',3'-Me$_2$, 36; 3'-MeO, 55; 2',3'-(MeO)$_2$, 3000). Unfortunately, we face the usual difficulty in proceeding from this knowledge of the transition state to a precise knowledge of the structure of the intermediate.

Rate data for the 7-norbornyl derivatives are summarized in Table 4.1.

The 3-phenyl-2-butyl system has been thoroughly explored.[52] The acetolysis of *threo-* and *erythro*-3-phenyl-2-butyl tosylates (**52**) proceeds with predominant retention of configuration. Consequently, it was proposed by Cram[53] that the reaction proceeded with the formation of a phenonium

[49]P. D. Bartlett and G. D. Sargent, *J. Amer. Chem. Soc.*, **87**, 1297 (1965).

[50]H. Goering and K. Humski, *J. Amer. Chem. Soc.*, **90**, 6213 (1968).

[51]H. Tanida, *Accounts Chem. Res.*, **1**, 239 (1968).

[52]C. J. Lancelot, D. J. Cram, and P. v. R. Schleyer, *Carbonium Ions*, Vol. III, G. A. Olah and P. v. R. Schleyer, Eds., Wiley–Interscience, New York, 1972, Chapter 27.

TABLE 4.1. Rates of Solvolysis of 7-Norbornyl Derivatives

7-Norbornyl derivative[m]	Leaving group X	Relative rates $10^6 k_1$, sec^{-1}	Relative rates		$\Delta H^{\ddagger}$, kcal mol^{-1}	$\Delta S^{\ddagger}$, eu
7-Norbornyl[a]	OTs	$6.36 \times 10^{-9\,b,c}$	1.00		35.7	-3.5
syn-7-Norbornenyl[d]	OTs	$2.6 \times 10^{-5\,b,c}$	10^4			
anti-7-Norbornenyl[a]	OTs	$904^{b,c}$	10^{11}		23.3	5.7
anti-7-Norbornenyl[e]	Cl	0.81^f				
7-Norbornadienyl[e]	Cl	612^f	$10^{14\,g}$			
anti-7-Norbornenyl[h]	OPNB	$4.72^{c,i}$	1.00		27.4	-17.3
2-methyl-anti-7-norbornenyl[h]	OPNB	$62.9^{c,i}$	13.0		27.8	-11.0
2,3-Dimethyl-anti-7-norbornenyl[h]	OPNB	$700^{c,i}$	150		27.2	-7.8
syn-7-Benzonorbornenyl[j]	OBs	$1.28 \times 10^{-2\,c,k}$		8.5×10^{-4}	31.6	-4.8
anti-7-Benzonorbornenyl[l]	OBs	$14.9^{c,k}$		1.00	27.7	-2.1
2'-Methoxy-anti-7-benzonorbornenyl[l]	OBs	$808^{c,k}$		54	28.9	9.3
2',3'-Dimethoxy-anti-7-benzonorbornenyl[l]	OBs	$44700^{c,k}$		3000		
2'-Methyl-anti-7-benzonorbornenyl[l]	OBs	$84.4^{c,k}$		5.7	27.4	-2.1
2',3'-Dimethyl-anti-7-benzonorbornenyl[l]	OBs	$535^{c,k}$		36.0		
2'-Chloro-anti-7-benzonorbornenyl[l]	OBs	$0.663^{c,k}$		0.045	32.2	4.7
2'-Nitro-anti-7-benzonorbornenyl[l]	OBs	$2.07 \times 10^{-3\,c,k}$		1.39×10^{-4}	33.2	-3.9

[a] Reference 42. [b] Rates are in acetic acid at 25°C. [c] Calculated from data at other temperatures. [d] S. Winstein and E. T. Stafford, J. Amer. Chem. Soc., **79**, 505 (1957). [e] Reference 43. [f] Rate in 80% acetone at 25°C. [g] Calculated by comparing the rates of anti-7-norbornenyl and 7-norbornadienyl chlorides (Reference 43). [h] Reference 48. [i] Rates in 70% dioxan at 25°C. [j] H. Tanida, Y. Hata, S. Ikegami, and Y. Ishitobi, J. Amer. Chem. Soc., **89**, 2928 (1967). [k] Rates in acetic acid at 77.6°C. [l] H. Tanida, T. Tsuji, and H. Ishitobi, J. Amer. Chem. Soc., **86**, 4904 (1964). [m] To retain relationship to the norbornyl system, the following numbering system is utilized:

bridge (**53**) which retained the stereochemistry and directed substitution to take place with retention (**54**) (27).

$$CH_3CH-CHCH_3 \longrightarrow CH_3CH-CHCH_3 \longrightarrow CH_3CHCHCH_3 \qquad (27)$$

OTs		OAc
52 *threo*	**53** *threo*	**54** *threo*

In the initial paper, Cram recognized the dilemma introduced by the failure to observe a significant rate enhancement over that of the parent compound, 2-butyl.[53] This was later discussed by Winstein, who suggested the possibility that stereochemical control could be achieved by rapid equilibration of open cations.[54]

This dilemma was later resolved by the understanding that solvent participation in the acetolysis of secondary tosylates can be large.[55]

By examining the rates and products of acetolysis of a number of *threo*-3-aryl-2-butyl brosylates containing representative substituents in the aromatic group (Table 4.2), it proved possible to separate the observed rate into a k_s term (the linear portion of the curve in Figure 4.2), and a k_Δ term. With these factors, it proved possible to realize a precise rate-product

[53]D. J. Cram, *J. Amer. Chem. Soc.*, **71**, 3863 (1949).
[54]S. Winstein and B. K. Morse, *J. Amer. Chem. Soc.*, **74**, 1133 (1952).
[55]P. v. R. Schleyer *et al.*, *J. Amer. Chem. Soc.*, **92**, 2538, 2540, 2542 (1970).

TABLE 4.2. Rates of Acetolysis of threo-3-Aryl-2-butyl Brosylates[56]

Aryl substituent	$10^5 k_1$, sec^{-1} at 75°C	Relative rate	$\Delta H^\ddagger$, kcal mol^{-1}	$\Delta S^\ddagger$, eu
p-CH$_3$O	1060	59		
p-CH$_3$	81.4	4.5	25.4	0.2
m-CH$_3$	28.2	1.6	26.4	0.8
H	18.0	1.0	26.6	0.4
p-Cl	4.53	0.25	27.9	1.3
m-Cl	2.05	0.11	27.1	−2.0
m-CF$_3$	1.38	0.078	27.0	−3.3
p-CF$_3$	1.26	0.070		
p-NO$_2$	0.495	0.028	29.0	−2.1
m,m'-(CF$_3$)$_2$	0.330	0.018		

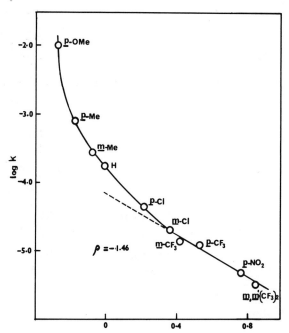

Figure 4.2. Rate of acetolysis of *threo*-3-aryl-2-butyl brosylates at 75.0°C vs. the σ-constants.

correlation, confirming aryl-participation and its effectiveness in controlling the stereochemistry of substitution.[56]

Solvent participation is far less important in the solvolysis of tertiary derivatives. Aryl participation is also less important. Accordingly, we undertook to synthesize a number of substituted 2,3-dimethyl-3-phenyl-2-butyl chlorides containing an appropriate tag and to examine their methanolysis products **(58, 59)**[57] (28).

The parent compound **(55)** (Z = H) shows 49% scrambling of the tag. The *p*-chloro derivative reveals even less: 19%. Scrambling increases with the introduction of activating groups: *p*-Me, 97%; *p*-CH$_3$O, 100%. Consequently, the results are consistent with an interpretation that in this system there is a rapid equilibration of the cation, **56 ⇌ 57**, whose rate increases in a predictable manner with the substituent in the aromatic ring. With *p*-methoxy, the rate of equilibration is too fast to trap it in this reaction before complete equilibration has occurred. In other reactions it has proved possible to trap this cation prior to full equilibration.[29]

[56]H. C. Brown and C. J. Kim, *J. Amer. Chem. Soc.*, **93**, 5765 (1971).
[57]H. C. Brown and C. J. Kim, unpublished research.

(28)

4.7. Conclusion

These results confirm that π- and aryl-participation, as revealed in rate studies (Chapter 3), lead to carbon-bridged intermediates. In other cases, where no significant participation is observed, the product can be a rapidly equilibrating pair or set of cations.

No authentic case of σ-participation was detected in the systems examined, and no evidence for σ-bridged cations was found.

We now turn our attention to cyclopropylcarbinyl (Chapter 5) and 2-norbornyl (Chapter 6). Do these provide authentic examples of σ-participation leading to σ-bridged nonclassical cations?

Comments

Stimulated by our attempts at arriving at a definition for "nonclassical" ions, many workers have offered their views, which reveal a rather broad spectrum of opinion. Some urge the replacement of the term altogether. Perhaps "Lewis" and "non-Lewis" are preferable alternatives to "classical" and "non-classical," but we favor retention of the designation which has been in use for a quarter of a century.

Except for traditional usage, the allyl cation, the cyclopropenium ion, and even benzene would be "nonclassical," not in the sense of our definition, but in the sense that none of these species can be adequately represented by a single Lewis structure. There is no difference in principle (except for π- vs. σ-character) between the three-center two-electron bonding present in the cyclopropenium ion and the bridged 2-norbornyl cation.

In contradiction to Professor Remick, we now recognize that resonance stabilization is possible in the methyl cation, $H-CH_2^+ \leftrightarrow H^+ CH_2$, and much of the positive charge resides on the hydrogen atoms. It is important not to think of the charge in classical carbocations as being "localized" and that in nonclassical ion as being "delocalized." Extensive delocalization is present in both instances, but the distribution of charge is, of course, different.

Present evidence indicates that, with the exception of the ethyl cation, simple primary carbocations, such as the n-propyl cation, are incapable of existence.[20] That is, they are not minima on a potential energy surface and are indefinable species, rearranging without activation to more stable structures. I do not believe that simple primary cations have been generated in solution. If intermediates exist under the conditions described,[25,26] they are not free carbocations, but are so highly encumbered by solvent and by leaving group as to change their character completely. Such "cationoid" or "carbonoid" species possess limited charge deficiency on carbon, and should not be considered to be carbocations. Many of the early experiments[5,25–27] which purported to exclude nonclassical ions in primary and secondary systems did not involve classical carbocations either.[37] In contrast, it is possible to generate essentially free *tertiary* carbocations in solvolytic media. Tertiary carbocations (e.g., **21**) are more stable, and have less to gain through bridging.

The ethyl cation is so unstable that it cannot be observed directly, even in superacid. The most sophisticated *ab initio* calculations employed to date on this problem predict the bridged structure (**6**) to be more stable than the classical (**5**) in the gas phase.[58] Similar calculations indicate the classical and

[58]B. Zurawski, R. Ahlrichs, and W. Kutzelnigg, *Chem. Phys. Lett.*, **21**, 309 (1973); J. A. Pople, private communication.

the bridged vinyl cation to be of comparable energy and both to be energy minima.[59]

Based on the large body of data which has been collected (Section 4.6), I find it hard to imagine ions like **60** and **61** as being other than symmetrically bridged. Analyses of hydrogen[60] (^{1}H) and carbon[61] (^{13}C) chemical shifts of **60**, its stable derivatives, and cyclopropylcarbinyl models provide strong additional supporting evidence. Perhaps quantum mechanical calculations will help resolve the exact nature of such species to Professor Brown's satisfaction.

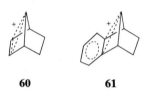

60   **61**

[*Comment added in proof* (HCB)] The complete summary of quantum mechanical calculations for the bridged and open structures of the ethyl cation, $C_2H_5^+$, does not lend confidence that we are yet at the stage where such calculations can provide a definitive answer.[62]

[*Reply* (PVRS)] The compilation referred to[62] is admittedly uncritical. Quite crude calculations are compared with the large basis set plus configuration interaction *ab initio* results cited in my *Comments*.[58,59] Calculations at the presently attainable limit on the vinyl cation isomers have been estimated to have uncertainties in relative energies no larger than ± 2 kcal/mole.[59]

[59]J. Weber and A. D. McLean, *J. Amer. Chem. Soc.*, **98**, 875 (1976); J. Weber, M. Yoshimine, and A. D. McLean, *J. Chem. Phys.*, **64**, 4159 (1976).

[60]R. K. Lustgarten, M. Brookhart, S. Winstein, P. G. Gassman, D. S. Patton, H. G. Richey, Jr., and J. D. Nichols, *Tetrahedron Lett.*, 1699 (1970); H. G. Richey, Jr., J. D. Nichols, P. G. Gassman, A. F. Fentiman, Jr., S. Winstein, M. Brookhart, and R. K. Lustgarten, *J. Amer. Chem. Soc.*, **92**, 3783 (1970).

[61]G. A. Olah and G. Liang, *J. Amer. Chem. Soc.*, **97**, 6803 (1975).

[62]B. Capon and S. P. McManus, *Neighboring Group Participation*, Vol. 1, Plenum Press, New York, 1976. See especially Figure 1 on p. 22 of this reference.

5

The Cyclopropylcarbinyl Cation

5.1. Introduction

The fascinating behavior of cyclopropylcarbinyl derivatives in solvolytic processes and other carbonium ion transformations[1] was explored in detail by J. D. Roberts and his coworkers.[2] The exceptionally fast rate of solvolysis and the facile interconversions of cyclopropylcarbinyl, cyclobutyl, and allylcarbinyl derivatives led to the proposal that σ-bridged intermediates were involved. Indeed, the first use of the term "nonclassical ion," appears to have been Roberts' proposed tricyclobutonium structure **2** for the cyclopropylcarbinyl cation **1**[3] (1).

$$\tag{1}$$

Fortunately, the cyclopropylcarbinyl cation possesses features which are especially helpful in clarifying the difference between hyperconjugation or σ-delocalization on the one hand, and σ-bridging on the other.

5.2. Exceptionally Fast Rates of Solvolysis

The rates of solvolysis of allyl and benzyl arenesulfonates are considerably faster than those of simple primary alkyl derivatives, such as ethyl,

---

[1] It has been possible in this chapter to cover only a small amount of pertinent data. For more detailed reviews, see H. G. Richey, Jr., *Carbonium Ions*, Vol. III, G. A. Olah and P. v. R. Schleyer, Eds., Wiley–Interscience, New York, 1972, Chapter 25; K. B. Wiberg, B. A. Hess, Jr., and A. J. Ashe III, *Carbonium Ions*, Vol. III, G. A. Olah and P. v. R. Schleyer, Eds., Wiley–Interscience, New York, 1972, Chapter 26.

[2] J. D. Roberts and R. H. Mazur, *J. Amer. Chem. Soc.*, **73**, 2509 (1951).

[3] J. D. Roberts and R. H. Mazur, *J. Amer. Chem. Soc.*, **73**, 3542 (1951).

presumably because of the resonance stabilization of the developing cationic center.[4] However, cyclopropylcarbinyl derivatives are even more reactive. Thus, the ethanolysis of cyclopropylcarbinyl benzenesulfonate (5) at 20°C is 14 times as rapid as that of allyl benzenesulfonate (4) and 500 times as rapid as that of ethyl benzenesulfonate (3)[5] (2). These rate differences are doubtless reduced by solvent participation.

$$
\begin{array}{cccc}
\text{CH}_2\text{O}_3\text{SPh} & \text{CH}_2\text{O}_3\text{SPh} & \text{CH}_2\text{O}_3\text{SPh} & \\
| & | & | & \\
\text{CH}_3 & \text{CH} & \triangle & \\
& \| & & \\
\mathbf{3} & \mathbf{4}\ \text{CH}_2 & \mathbf{5} & \tag{2}
\end{array}
$$

RR (20°C):   1.00   36   500

For example, such solvent participation is much smaller in tertiary derivatives. Here the solvolysis in 80% acetone reveals major increases from the *tert*-butyl system (6), to the *tert*-cumyl system (7), to the *tert*-cyclopropyldimethylcarbinyl derivative (8)[6] (3).

$$
\begin{array}{ccc}
\text{H}_3\text{C} & \text{CH}_3 & \text{CH}_3 \\
| & | & | \\
\text{H}_3\text{C}-\overset{}{\text{C}}-\text{OPNB} & \text{H}_3\text{C}-\overset{}{\text{C}}-\text{OPNB} & \text{H}_3\text{C}-\overset{}{\text{C}}-\text{OPNB} \\
\text{H}_3\text{C} & & \\
\mathbf{6} & \mathbf{7} & \mathbf{8}
\end{array} \tag{3}
$$

RR (25°C):   1.00   969   503,000

Likewise, Hart revealed the enormous stabilizing effect of a cyclopropylcarbinyl group relative to an alkyl group, such as isopropyl[7] (4).

$$
\begin{array}{cccc}
\rangle\!-\!\text{C}-\text{X} & \rangle\!-\!\text{C}-\text{X} & \rangle\!-\!\text{C}-\text{X} & \triangleright\!-\!\text{C}-\text{X}
\end{array} \tag{4}
$$

RR (25°C):   1.00   250   25,000   25,000,000

A detailed study has been made of the rates of solvolysis (and products) in methanol and acetic acid of cyclopropylcarbinyl β-naphthalenesulfonate (ROβNs) and other representative primary and secondary derivatives.[8] (The β-naphthalenesulfonates were adopted because the derivatives were crystalline and a crystalline cyclopropylcarbinyl ester assured its homogeneity and purity.)

[4]A. Streitwieser, Jr., *Solvolytic Displacement Reactions*, McGraw-Hill, New York, 1962.
[5]C. G. Bergstrom and S. Siegel, *J. Amer. Chem. Soc.*, **74**, 145, 254 (1952).
[6]H. C. Brown and E. N. Peters, *J. Amer. Chem. Soc.*, **95**, 2400 (1973).
[7]H. Hart and P. A. Law, *J. Amer. Chem. Soc.*, **86**, 1957 (1964); H. Hart and J. M. Sandri, *J. Amer. Chem. Soc.*, **81**, 320 (1959).
[8]H. C. Brown and S. Nishida, manuscript in preparation.

In methanol, cyclopropylcarbinyl is slightly more reactive than benzyl, and both are enormously more reactive than ethyl- and allylcarbinyl (5). In acetic acid there is a major enhancement of rate for the cyclopropylcarbinyl derivatives.

$$\underset{\text{CH}_3}{\overset{\text{CH}_2O\beta\text{Ns}}{|}} \qquad \underset{\text{CH}_2}{\overset{\text{CH}_2O\beta\text{Ns}}{\underset{\overset{|}{\text{CH}}}{|}}} \qquad \text{CH}_2O\beta\text{Ns} \qquad \text{CH}_2O\beta\text{Ns} \tag{5}$$

RR (25°C):

Methanol	1.00	0.278	788	1,220
Acetic acid	1.00	0.502	1,880	139,000

The cyclobutyl derivative also reveals a significant rate enhancement over other secondary derivatives in acetic acid as compared with methanol (6).

$$\underset{H_3C \diagup \overset{|}{\underset{\diagdown}{CH}} \diagdown CH_3}{} \qquad O\beta\text{Ns} \qquad O\beta\text{Ns} \qquad O\beta\text{Ns} \qquad O\beta\text{Ns} \tag{6}$$

RR (25°C):

Methanol	1.00	1.22	8.22	0.153
Acetic acid	1.00	19.6	23.7	0.665

These data are summarized in Table 5.1.

TABLE 5.1. *Relative Rates of Solvolysis of Representative β-Naphthalenesulfonates (ROβNs)*

	Methanolysis, at 25°C		Acetolysis, at 25°C	
β-Naphthalenesulfonate	$10^6 k_1$, sec^{-1}	Rel. rate[a]	$10^6 k_1$, sec^{-1}	Rel. rate[a]
EthylOβNs	0.439	1.00	0.00285[b,c]	1.00
IsobutylOβNs[c]	0.0150	0.0342	0.00035[b,c]	0.123
AllylcarbinylOβNs	0.122	0.278	0.00143	0.502
CyclopropylcarbinylOβNs	536	1220	397	139000
BenzylOβNs	346	788	5.37	1880
IsopropylOβNs	2.02	1.00	0.110[c]	1.00
CyclobutylOβNs	2.47	1.22	2.16	19.6
CyclopentylOβNs	16.6	8.22	2.61	23.7
CyclohexylOβNs	0.310	0.153	0.00731	0.665

[a] Relative to ethyl for the primaries and to isopropyl for the secondaries.
[b] Calculated from data for the tosylate, S. Winstein and H. Marshall, *J. Amer. Chem. Soc.*, **74**, 1120 (1952), using the factor ROβNs/ROTs = 1.6.
[c] Calculated from data at higher temperatures. Calculated from data for the tosylate, S. Winstein, E. Grunwald, and H. W. Jones, *J. Amer. Chem. Soc.*, **73**, 2700 (1951).

It is evident that the rates of solvolysis of cyclopropylcarbinyl derivatives are enormously enhanced. The rates for cyclobutyl appear less
enhanced. However, if one corrects for the rate-retarding effect of I-strain,[9]
then it also must be considered to exhibit an enhanced rate. The question to
be settled is whether these rate enhancements are to be attributed to
σ-participation leading to the formation of a σ-bridged nonclassical cation,
such as **2**.

5.3. Rearrangements in the Solvolysis of Cyclopropylcarbinyl Derivatives

The solvolysis of both cyclopropylcarbinyl and cyclobutyl derivatives
proceeds with rearrangement to give mixtures of cyclopropylcarbinyl, cyclobutyl, and allylcarbinyl solvolysis products.[2,3] Under certain conditions,
such as the solvolysis of chlorides in acetic acid or deamination of the amines
with aqueous nitrous acid, very similar mixtures of products are obtained
from both cyclopropylcarbinyl and cyclobutyl.[2,3]
We undertook to examine the solvolysis of cyclopropylcarbinyl and
cyclobutylOβNs derivatives in acetic acid and methanol, both in the presence and absence of base. In acetic acid, the reaction is complicated by major
internal return, resulting in the formation of approximately 45% of cy-
clobutylOβNs with a small amount ($\sim$5%) of allylcarbinylOβNs. Since the
rates of solvolysis of these derivatives is far slower than that of cyclopropylcarbinyl (Table 5.1), it is possible to examine the products produced in the
solvolysis of this reactive species.[7] The normalized results are shown in (7).
Clearly the solvolysis products from both systems are very similar.

$$\triangleright\!\!-CH_2O\beta Ns \xrightarrow[\text{AcOH}]{25°C} \triangleright\!\!-CH_2OAc \qquad \square\!\!-OAc \qquad \begin{array}{c} CH_2=CH \\ | \\ CH_2CH_2OAc \end{array}$$

0.1 M

AcOH	51%	56	37	7
AcONa (0.11 M)	53%	58	36	5

$$\square\!\!-O\beta Ns \xrightarrow[\text{AcOH}]{25°C}$$

0.1 M

AcOH		53	40	7
AcONa (0.11 M)	93%	52	36	6

(7)

The results in methanol are quite different. First, internal return is a less
significant problem, with solvolysis readily proceeding to 93% completion
for cyclopropylcarbinyl and to 100% completion for cyclobutyl. Secondly,

[9]H. C. Brown and M. Borkowski, *J. Amer. Chem. Soc.*, **74**, 1894 (1952).

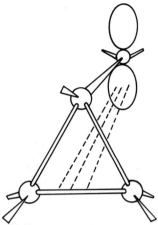

Figure 5.1. σ-Bridging leading to the proposed tricyclobutonium cation. From Streitwieser.[4]

the products obtained from the two systems reveal considerable differences (8).

$$\triangleright\!\!-CH_2O\beta Ns \xrightarrow[\underset{0.1\,M}{CH_3OH}]{25°C} \triangleright\!\!-CH_2OCH_3 \quad \langle\!\rangle\!\!-OCH_3 \quad CH_2=CH\!\!-\!\!CH_2CH_2OCH_3$$

CH_3OH	93%	66	31	3
CH_3ONa (0.11 M)	94%	69	28	2

$$\langle\!\rangle\!\!-O\beta Ns \xrightarrow[\underset{0.1\,M}{CH_3OH}]{25°C} \tag{8}$$

CH_3OH	100%	48	47	5
CH_3ONa (0.11 M)	100%	47	47	6

The results are not compatible with the production of a single σ-bridged intermediate, such as **2**. It is compatible with the production of a rapidly equilibrating set of cyclopropylcarbinyl ⇌ cyclobutyl cations with a small, essentially irreversible, diversion to the allylcarbinyl derivatives.[10]

5.4. σ-Bridged Structures for the Cation

The remarkably enhanced rates of solvolysis of cyclopropylcarbinyl derivatives and the similarity in the products from carbonium ion reactions of cyclopropylcarbinyl and cyclobutyl derivatives[1] led to the proposal that the reaction proceeds with σ-participation to produce the symmetrical tricyclobutonium ion intermediate[3] (**2**).

The proposal is made clear in Figure 5.1 from Streitwieser.[4]

[10] Z. Majerski, S. Borčić, and D. E. Sunko, *Tetrahedron*, **25**, 301 (1969), discuss an alternative interpretation of similar results.

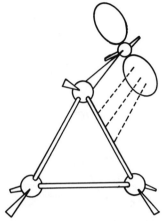

Figure 5.2. σ-Bridging leading to the proposed bicyclobutonium cation. From Streitwieser.[4]

Later it was observed that the reactions of tagged cyclopropylcarbinyl derivatives fail to show the full equilibration of the tag required by the tricyclobutonium ion.[11] It was therefore proposed that the cyclopropylcarbinyl cation exists instead as a rapidly equilibrating set of six equivalent bicyclobutonium cations (9).

$$\triangle\!\!\!\rangle \;\rightleftharpoons\; \triangle\!\!\!\wedge \;\rightleftharpoons\; \triangle\!\!\!\wedge \quad \text{etc.} \tag{9}$$

The proposed σ-participation with the formation of a σ-bridged intermediate is again made clear by the diagram (Figure 5.2) from Streitwieser.[4]

The relationship to a set of rapidly equilibrating cyclopropylcarbinyl cations (10) is apparent. All that would be required to account for the

$$\triangle\!\!\!\rangle \;\rightleftharpoons\; \triangle\!\!\!\wedge \;\rightleftharpoons\; \triangle\!\!\!\wedge \tag{10}$$

products is to postulate fast equilibration with a small concentration of a puckered cyclobutyl cation (9).[12,13]

9

Indeed, the interconversion of the cyclopropylcarbinyl cations (10) may proceed through such a puckered cyclobutyl cation.[1]

Fortunately, the stereochemical characteristics of the cyclopropylcarbinyl cation make it possible to rule out σ-bridged cations as a factor in these reactions.

[11]R. H. Mazur, W. N. White, D. A. Semenov, C. C. Lee, M. S. Silver, and J. D. Roberts, *J. Amer. Chem. Soc.*, **81**, 4390 (1959).

[12]K. B. Wiberg and G. Szeimies, *J. Amer. Chem. Soc.*, **92**, 571 (1970).

[13]Z. Majerski and P. v. R. Schleyer, *J. Amer. Chem. Soc.*, **93**, 665 (1971).

5.5. Stereochemical Characteristics

There is now much evidence that major electron supply from the cyclopropyl group can occur in systems where σ-bridging cannot take place for stereochemical reasons. For example, 3-nortricyclyl derivatives are enormously more reactive than 7-norbornyl (11).[14,15] Yet the developing

RR (25°C):   1.00 10^8 (11)

carbonium ion center is in the bisected arrangement which is stereochemically in an unfavorable position to form a σ-bridge with the cyclopropane ring.[16]

On the other hand, when the cyclopropyl ring is forced into the parallel arrangement most favorable for σ-bridging, no rate enhancement is observed (12).[17]

RR (45°C):   1   2.3   0.0065   (12)

The cyclopropyl group can supply electrons to stabilize an electron-deficient center even when remote from the carbonium ion center (13).[18]

RR (25°C):   157   20.7   23.7

[14]H. G. Richey, Jr., and N. C. Buckley, *J. Amer. Chem. Soc.*, **85**, 3057 (1963).
[15]H. C. Brown and E. N. Peters, *J. Amer. Chem. Soc.*, **97**, 1927 (1975).
[16]J. D. Roberts, W. Bennett, and R. Armstrong, *J. Amer. Chem. Soc.*, **72**, 3329 (1950).
[17]B. R. Ree and J. C. Martin, *J. Amer. Chem. Soc.*, **92**, 1660 (1970); V. Buss, R. Gleiter, and P. v. R. Schleyer, *ibid.*, **93**, 3927 (1971).
[18]R. C. Hahn, T. F. Corbin, and H. Shechter, *J. Amer. Chem. Soc.*, **90**, 3404 (1968).

Its electronic effects are sensitive to conformation, as revealed by the results shown in (14).[19] If we correct for the rate enhancing effect of

(14)

RR (25°C):	157	172	37
(corrected)	(157)	(86)	(9)

m-methyl ($\times$2), the corrected values in parentheses are obtained. Note that *p*-isopropyl shows a rate-enhancing effect of 18 over *p*-hydrogen (1.00). Consequently, the *p*-cyclopropyl substituent forced out of the bisected conformation by the *m*-methyl substituents into the parallel arrangement is even less effective than the isopropyl group in delocalizing charge.

These results are summarized in Table 5.2.

The nmr spectrum of the dimethylcyclopropylcarbinyl cation (**10**) reveals it to exist in the bisected conformation with one methyl group *cis* and the other *trans* to the cyclopropyl ring.[20] The barrier to rotation is

10

13.7 kcal mol^{-1}.[21] Consequently, there is little doubt that the stabilization of the carbonium ion center is best in the bisected conformation.

Equally convincing are the results of a study of the solvolysis of optically active methylcyclopropylcarbinyl derivatives. Here the formation of a σ-bridged ion would be expected to protect the bridged side of the carbonium ion, resulting in retention of configuration in the product. However, the solvolysis product is inactive.[22]

Consequently, there is now a huge mass of data to support the position that electron contributions from the cyclopropyl ring to the electron-deficient center occurs best from the bisected arrangement (**11**).

11   **12**

[19]H. C. Brown and J. D. Cleveland, *J. Amer. Chem. Soc.*, **88**, 2051 (1966).
[20]G. A. Olah, C. L. Jeuell, D. P. Kelly, and R. D. Porter, *J. Amer. Chem. Soc.*, **94**, 146 (1972).
[21]D. S. Kabakoff and E. Namanworth, *J. Amer. Chem. Soc.*, **92**, 3234 (1970).
[22]M. Vogel and J. D. Roberts, *J. Amer. Chem. Soc.*, **88**, 2262 (1966).

TABLE 5.2. *Effect of Substituents and Conformation in the Cyclopropyl System on the Rates of Solvolysis*

Cyclopropyl derivative	Leaving group X	$10^6 k_1$, sec^{-1}	— Relative rates —	
Cyclopropylcarbinyl[a]	ODNB	0.43[b]	1.00	
1-Methylcyclopropylcarbinyl[a]	ODNB	2.13[b]	5.0	
trans-2-Methylcyclopropylcarbinyl[a]	ODNB	4.75[b]	11.0	
cis-2-Methylcyclopropylcarbinyl[a]	ODNB	3.50[b]	8.2	
2,2-Dimethylcyclopropylcarbinyl[a]	ODNB	39.7[b]	92	
trans,trans-2,3-Dimethylcyclo- propylcarbinyl[a]	ODNB	53.3[b]	124	
cis,cis-2,3-Dimethylcyclo- propylcarbinyl[a]	ODNB	35.3[b]	82	
cis,trans-2,3-Dimethylcyclo- propylcarbinyl[a]	ODNB	34.5[b]	80	
trans-2,3,3-Trimethylcyclo- propylcarbinyl[a]	ODNB	212[b]	490	
2,2,3,3-Tetramethylcyclo- propylcarbinyl[a]	ODNB	675[b]	1570	
trans-2-Ethoxycyclopropylcarbinyl[a]	ODNB	403[b]	940	
1-Adamantyl[c]	OTs	7430[d]	1.00	
2,2-Dimethyl-1-adamantyl[c]	OTs	1.91×10^4 [d]	2.3	
2-Cyclopropyl-1-adamantyl[c]	OTs	50.3[d]	0.0065	
tert-Cumyl[e]	Cl	124[f]		1.00
4-Isopropyl-tert-cumyl[e]	Cl	2210[f]		17.8
3-Methyl-tert-cumyl[e]	Cl	248[f]		2.0
4-Cyclopropyl-tert-cumyl[e]	Cl	1.95×10^4 [f,g]		157
3-Methyl-4-cyclopropyl-tert- cumyl[e]	Cl	2.13×10^4 [f,g]		172
3,5-Dimethyl-tert-cumyl[e]	Cl	473[f]		3.9
3,5-Dimethyl-4-cyclopropyl-tert- cumyl[e]	Cl	4600[f,g]		37.1

[a]Reference 24. [b]Rates in 60% aqueous acetone at 100°C. [c]Reference 17. [d]Rates in acetic acid at 45°C. [e]Reference 19. [f]Rates in 90% acetone at 25°C. [g]Rates extrapolated from data at lower temperatures.

The σ-bridged structure would require a rotation of the carbonium carbon away from this stabilized conformation toward the parallel arrangement **12** where one lobe of the p-orbital is directed toward the cyclopropane ring.

The data we have discussed have been based on studies with secondary and tertiary cyclopropylcarbinyl cations. Recently, it was proposed, on the basis of nmr studies, that the primary cyclopropylcarbinyl cation, in contrast to the secondary and tertiary, must be σ-bridged.[20,23]

[23]G. A. Olah, D. P. Kelly, C. L. Jeuell, and R. D. Porter, *J. Amer. Chem. Soc.*, **92**, 2544 (1970).

However, study of the solvolysis rates of methyl substituted cyclo-propylcarbinyl dinitrobenzoates reveal reasonably good additivity with the

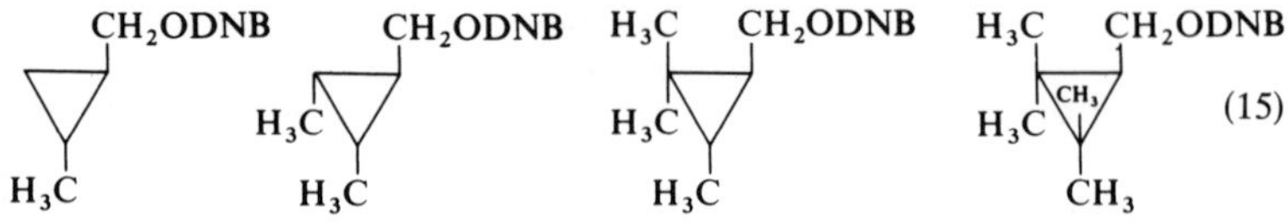

(15)

RR (100°C):   11   124   490   1570
(Unsubst. = 1.00)

introduction of one, two, three, or four methyl groups (15).[24] A carbon-bridged structure for the tetramethyl derivative would require a $\pi\sigma$-bridge to a highly substituted, highly congested steric environment (**13**).

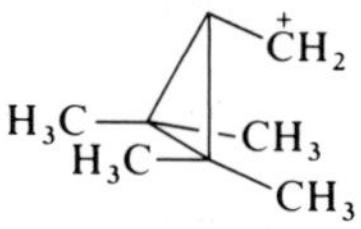

13

Recent *ab initio* calculations also support the existence of the cyclo-propylcarbinyl cation as a rapidly equilibrating set of open cations in the bisected arrangement.[25]

Finally, an optically active primary cyclopropylcarbinyl derivative was prepared and solvolyzed.[26] Again no evidence for retention, anticipated for a σ-bridged structure, was observed.

5.6. Conclusion

These results appear to rule out σ-participation and σ-bridging in the cyclopropylcarbinyl system, except possibly for the primary system under the special conditions of superacid solution.[20] However, as discussed later (Chapter 13), even here the situation is in doubt. Consequently, let us turn our attention to 2-norbornyl, considered to be the optimum case for σ-bridging.

[24]P. v. R. Schleyer and G. W. Van Dine, *J. Amer. Chem. Soc.*, **88**, 2321 (1966).

[25]W. J. Hehre and P. C. Hiberty, *J. Amer. Chem. Soc.*, **96**, 302 (1974); W. J. Hehre, *Accounts Chem. Res.*, **8**, 369 (1975).

[26]C. D. Poulter and C. J. Spillner, *J. Amer. Chem. Soc.*, **96**, 7591 (1974).

Comments

Brown's chapter concentrates on arguments against tricyclobutonium (**2**; Figure 5.1) and bicyclobutonium [**14**; Figure 5.2; (**9**)] formulations for the cyclopropylcarbinyl cations; a third major alternative, the homoallylic representation **15** is not mentioned specifically. All of these species imply a different charge distribution and have been discussed widely in the literature.[1,24]

2	**14**	**15**

Professor Brown and I agree that the available data indicate that the parent cyclopropylcarbinyl cation prefers the bisected conformation rather than structures implied by **2**, **14**, or **15**. We disagree, however, whether this species should be considered to be "classical" or "nonclassical."

My 1966[24] viewpoint remains appropriate, "Many structures for the cyclopropylcarbinylcarbonium ion have been considered. The classical representation, **1**, implies to us localization of the charge to the 1' position in a manner similar to that in the isobutyl or in the cyclohexylcarbinyl cation. We regard **1** as unsatisfactory in view of the abundant evidence for conjugation; some kind of 'dotted line' formulation appears to us preferable." Structure **16** has most commonly been used to represent the nonclassical cyclopropylcarbinyl cation in the *bisected conformation*. How can these "dotted lines" be justified?

1	**16**

The allyl cation affords an excellent analogy. The less stable perpendicular conformation is represented by structure **17** reasonably well. (The unequal C—C bond lengths given are taken from *ab initio* calculations.)[27] The planar conformation of the allyl cation (**18**) is about 35 kcal/mole more stable than **17** and the C—C bond lengths are equal (1.385 Å).[27] Clearly, the single Lewis structure **19** does not provide an adequate description of this species; formulation **18** with its dotted lines is preferable.

[27]L. Radom, P. C. Hariharan, J. A. Pople, and P. v. R. Schleyer, *J. Amer. Chem. Soc.*, **95**, 6531 (1973).

17   **18**   **19**

The STO-3G *ab initio* geometries[25] of the perpendicular form **20** (a hypothetical species) and the bisected form **21** of the cyclopropylcarbinyl cation show variations as great as those differentiating **17** and **18**. Although the perpendicular cyclopropylcarbinyl cation (**20**) is well represented by a single Lewis structure (e.g., **1**), the marked bond length changes, the more extensive charge delocalization and the much greater stability (26 kcal/mole) of the bisected form justify the "dotted line" formulation (**16, 21**). It does not seem reasonable to me to represent both **20** and **21** by the same formula (**1**) when the properties are as different as the two conformations (**17** and **18**) of the allyl cation.

20   **21**   **22**
perpendicular   bisected

The present understanding is that the cyclopropylcarbinyl cation does not possess the σ-bridged nonclassical structures **2** or **14**. Should the bisected cyclopropylcarbinyl cations (**16, 21**) be considered to be "nonclassical"? Overlooking the historical justification for the use of that term,[3] does it fit our proposed definition (Chapter 4.2)? For reasons discussed, I do not think **16** or **21** can *adequately* be represented "by a single Lewis structure." The vinyl group in **16, 21** can be regarded to be bridging in much the same way as the methyl in **22**. Allyl carbinyl precursors (**23**) can be used to generate cyclopropylcarbinyl cations although they are usually less well suited because they are prone to competing processes.[1] The bridging carbon, C1 in **16** or **21**, has increased its coordination number from *three* in **23** to *four* in **16** or **21**. However, if cyclopropylcarbinyl derivatives (**24**) are the ion precursors, no increase in coordination number occurs. As structure **21** indicates, the characteristics of three-center bonding, i.e., long C-1,2 and C-1,3 bonds, are present in this species.

23   **16**   **24**

Hyperconjugation of *normal* magnitude is present in **20**,[28] but is enormously enhanced in **21**. However, hyperconjugation alone, without geometry alteration, accounts for only part of the stabilizing effect. Thus, *rigid rotation* of the $C-\overset{+}{C}H_2$ bond in the *fixed geometry* cyclopropylcarbinyl cations requires only 17.5 kcal/mole (STO-3G), whereas fully optimized **20** and **21** differ by 26 kcal/mole.[25] Thus, geometry relaxation (bond lengthening) provides 8.5 kcal/mole extra stabilization. The barrier to rotation in the dimethylcyclopropylcarbinyl cation, although diminished, remains large, 13.7 kcal/mole (exptl.)[21] and 16.6 kcal/mole (calc.).[25]

With regard to our definition and with regard to the opinion expressed to me by a number of workers, the cyclopropylcarbinyl cation is a borderline case with regard to its designation as "classical" or "nonclassical." All agree that it is a unique species with properties quite different from those expected of simple primary aliphatic cations.

Although the tendency of cyclopropylcarbinyl cations to adopt bisected or nearly bisected geometries is quite large, rigid systems are known where this is impossible.[17] The series (16) discussed by Rhodes and DiFate[29] is particularly instructive; the estimated rate differences based on comparisons with model compounds lacking the cyclopropane rings is given beneath each structure. The estimated dihedral angles are also given.[29]

(16)

RR (25°C):	2.5×10^8	4.0×10^5	4.1×10^3	5.0×10^{-3}
Parent[29] = 1.00	bisected			perpendicular

Dihedral angles	~0°	~30°	~60°	~90°

The ion from **25**, an intermediate case, probably is best represented by structure **26**. Should this species be called a "homoallylic ion" or is it best to regard it simply as a partially rotated cyclopropylcarbinyl cation? Chemists will differ in their opinion.

26

[28]L. Radom, J. A. Pople, and P. v. R. Schleyer, *J. Amer. Chem. Soc.*, **94**, 5935 (1972).

[29]Y. E. Rhodes and V. G. DiFate, *J. Amer. Chem. Soc.*, **94**, 7582 (1972). The parent is the corresponding structure without the cyclopropane moiety.

Finally, there is a problem in Brown's rationalization of the rapid equilibration among the "classical" cyclopropylcarbinyl cations, **27–29**:

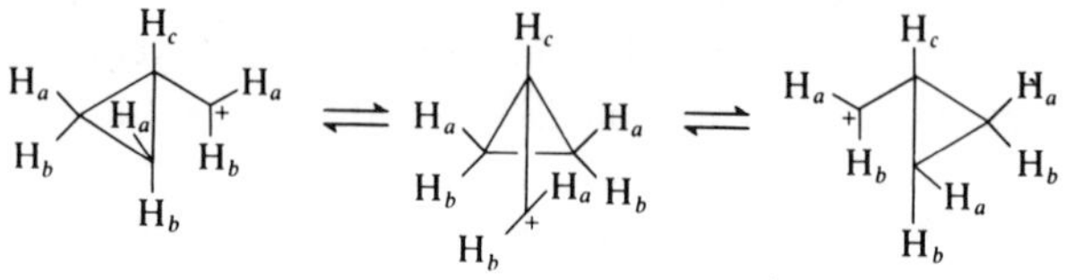

Both in solvolytic media[12,13] and in superacid,[20,23] protons H_a, H_b, and H_c do not undergo any site exchanges during degenerate rearrangement. If a puckered cyclobutyl cation (**30**) is involved as intermediate, the barrier to ring inversion **30** $\rightleftharpoons$ **31** must be large or else protons H_a and H_b would become equivalent.[25] Such a large barrier is not consistent with the behavior I expect of a classical cyclobutyl cation. Thus, cyclobutane itself has a ring-inversion barrier of only 1.4 kcal/mole,[30] cyclobutanone, with an sp^2 center in the ring, is planar,[31] and the classical 1-phenylcyclobutyl cation is planar at $-65°$, at least on the time-average nmr scale.[20] Some sort of "nonclassical" intermediate or bridged transition state would appear to be involved in the **27** $\rightleftharpoons$ **28** $\rightleftharpoons$ **29** process; **32** has been proposed on the basis of *ab initio* calculations.[25]

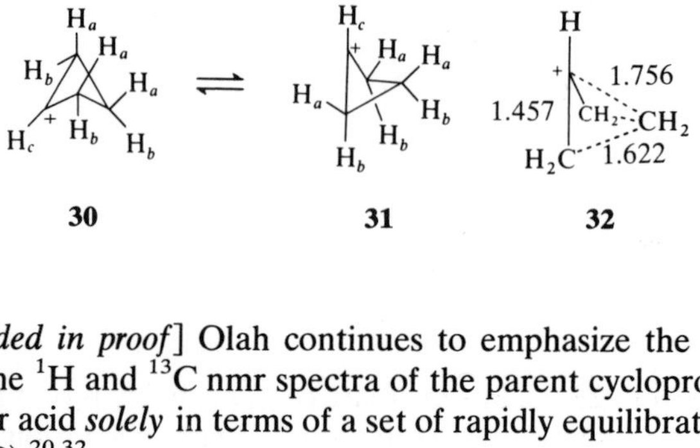

[*Note added in proof*] Olah continues to emphasize the difficulty of interpreting the ^1H and ^{13}C nmr spectra of the parent cyclopropylcarbinyl cation in super acid *solely* in terms of a set of rapidly equilibrating bisected cations (**27–29**).[20,32]

[30]J. M. R. Stone and I. M. Mills, *Mol. Phys.*, **18**, 631 (1970).
[31]L. H. Scharpen and V. W. Laurie, *J. Chem. Phys.*, **49**, 221 (1968).
[32]G. A. Olah, R. J. Spear, P. C. Hiberty, and W. J. Hehre, *J. Amer. Chem. Soc.*, **98**, 7470 (1976).

6

The 2-Norbornyl Cation

6.1. Introduction

At the time we undertook our examination of the validity of the proposals that σ-bridging represents a major factor in the rates of solvolysis, there was general acceptance for the position that cyclopropylcarbinyl (Chapter 5) and 2-norbornyl[1-4] provided the most favorable examples for this phenomenon.

Fortunately, it is now clear that the special stereochemical requirements for electron supply from the cyclopropyl group to the electron-deficient center are in direct opposition to those required for a σ-bridge (Figures 5.1 and 5.2). This feature greatly facilitated the task of reaching a decision for the cyclopropylcarbinyl cation (Chapter 5). At this time there appears to be little remaining support for the significance of a σ-bridged species in the solvolysis of cyclopropylcarbinyl derivatives.[5,6]

2-Norbornyl has provided a much more difficult case to resolve. In this case the stereochemical requirements for a σ-bridge, even though not ideal, are reasonably met by the bicyclo[2.2.1]heptane structure. Consequently, it

[1] S. Winstein and D. Trifan, *J. Amer. Chem. Soc.*, **74**, 1147, 1154 (1952).
[2] P. D. Bartlett, *Nonclassical Ions*, Benjamin, New York, 1965.
[3] J. A. Berson, *Molecular Rearrangements*, Vol. 1, P. de Mayo, Ed., Interscience, New York, 1963, Chapter 3.
[4] G. D. Sargent, *Carbonium Ions*, Vol. III, G. A. Olah and P. v. R. Schleyer, Eds., Wiley–Interscience, New York, 1972, Chapter 24.
[5] H. G. Richey, Jr., *Carbonium Ions*, Vol. III, G. A. Olah and P. v. R. Schleyer, Eds., Wiley–Interscience, New York, 1972, Chapter 25.
[6] K. B. Wiberg, B. A. Hess, and A. J. Ashe III, *Carbonium Ions*, Vol. III, G. A. Olah and P. v. R. Schleyer, Eds., Wiley–Interscience, New York, 1972, Chapter 26.

has been necessary to employ more subtle probes. The problem is a fascinating one, worthy of the finest intellectual effort. Those who approach the question objectively will enjoy a most interesting, stimulating experience.

6.2. Nonclassical "Structures" for the 2-Norbornyl Cation

Numerous "structures" have been proposed for 2-norbornyl. It is indeed rare for a single species to be so honored. It is appropriate here to summarize the situation before considering in detail the experimental facts that the structures attempt to interpret.

First, there is the classical structure (1). As discussed earlier (Section 4.3), this structure will entail electronic readjustments under the influence of the electron deficiency, electronic readjustments which will reflect inductive, inductomeric, field and hyperconjugative influences.

Clearly this cannot be the structure of the 2-norbornyl cation. It does not possess a plane of symmetry and should therefore be formed and transformed into optically active products. Solvolysis of the optically active brosylate leads to inactive product. A rapidly equilibrating pair of ions or ion pairs (2) possess such a plane of symmetry and constitutes a possible structure for the 2-norbornyl intermediate.[1]

Winstein proposed that the transition state (3) for such an equilibrating pair might be sufficiently stable so as to replace the classical structure.[1] He

represented this structure 3 as a resonance hybrid of three canonical structures (3'-3''').

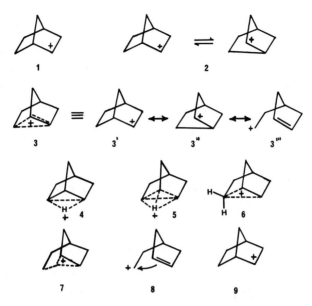

Figure 6.1. Proposed structures of the 2-norbornyl cation.

He also considered the edge-protonated species[1] (**4**), while Roberts proposed the face-centered species (**5**).[7] On the other hand, Olah has favored a formulation as a corner-protonated nortriclene (**6**),[8] deleting the dashed double-bond of Winstein in (**3**). More recently, he has favored still another formulation (**7**).[9] Dewar prefers the formulation as a π-complex (**8**).[10] Finally, both Jensen and Traylor have supported a formulation of the

2-norbornyl cation as involving vertical stabilization by the C_1–C_6 bonding pair, but without σ-bridging or movement of the atoms (**9**).[11,12] In other words, the structure **9** would be that of the classical ion **1**, but stabilized by an electronic contribution that does not alter the geometry. (I find it difficult to

[7] J. D. Roberts, C. C. Lee, and W. H. Saunders, Jr., *J. Amer. Chem. Soc.*, **76**, 4501 (1954).
[8] G. A. Olah and A. M. White, *J. Amer. Chem. Soc.*, **91**, 5801 (1969).
[9] G. A. Olah and A. M. White, *J. Amer. Chem. Soc.*, **94**, 808 (1972).
[10] M. J. S. Dewar and A. P. Marchand, *Ann. Rev. Phys. Chem.*, **16**, 321 (1965).
[11] F. R. Jensen and B. E. Smart, *J. Amer. Chem. Soc.*, **91**, 5686, 5688 (1969).
[12] T. G. Traylor, W. Hanstein, H. J. Berwin, N. A. Clinton, and R. S. Brown, *J. Amer. Chem. Soc.*, **93**, 5715 (1971).

understand why Traylor calls this a "nonclassical ion." It does not possess the symmetry characteristic of the nonclassical formulations. The classical formulation would include hyperconjugative interactions in the ion **1**, but would permit the usual minor readjustments in the structure accompanying such interactions.)

These nonclassical structures are summed up in Figure 6.1.

6.3. Basic Facts

What are the experimental bases for the proposed σ-bridged structure of the 2-norbornyl cation?

A. Exceptionally Fast Rates for Exo Derivatives

Ingold[13] observed that the rate of solvolysis of camphene hydrochloride (**10**) was 6000 (actually 13,600)[14] times that of *tert*-butyl chloride (1). He

$$\text{10} \qquad\qquad \text{11} \tag{1}$$

argued that in his opinion such a fast rate could not arise from simple relief of steric strain (Chapter 2). Consequently, he proposed the formation of a stabilized "synartetic" ion **11**.

B. High Exo:Endo Rate Ratios

Winstein and Trifan pointed out that *exo*-norbornyl brosylate (**12**) undergoes acetolysis (2) at a rate 350 times greater than that of the *endo* isomer (**13**).[1] It was postulated that the 1,6-bonding pair is in an ideal

$$\text{12} \qquad \text{(transition state)} \qquad \text{3} \tag{2}$$

[13]F. Brown, E. D. Hughes, C. K. Ingold, and J. F. Smith, *Nature*, **168**, 65 (1951).
[14]H. C. Brown and F. J. Chloupek, *J. Amer. Chem. Soc.*, **85**, 2322 (1963).

position to participate in the displacement of the *exo*-OBs group (**12**), but it is in an unfavorable position to assist the ionization of the *endo* isomer **13**.

13

C. Racemization

Solvolysis of optically active *exo*-norbornyl brosylate in acetic acid, 80% dioxane, and 75% acetone at 25°C yields racemized ($\geq$99.95%) *exo*-norbornyl derivatives.[1,15] Solvolysis of optically active *endo*-norbornyl brosylate also yields *exo*-norbornyl products, but with a small amount of retained activity[1,15] (13% in 75% acetone, 7% in acetic acid, and 3% in formic acid). This product must involve solvent attack with inversion on the *endo*-brosylate, or the corresponding ion pair.

The rate of racemization of the *exo*-isomer exceeds the titrimetric rate of solvolysis by factors of 1.40 in 75% acetone, to 2.94 in ethanol, to 3.46 in acetic acid[1] (later revised to 4.6, but without supporting data[15]). This is attributed to recapture of the anion by the cation from the first formed σ-bridged or rapidly equilibrating ion pair to reform racemized brosylate. If the rate of racemization of the *exo* isomer, k_α, is taken as equal to the rate of ionization, then the *exo*:*endo* rate ratio becomes 1600.[15]

Variable amounts of 6:2 and 3:2 hydride shifts occur during solvolysis and other carbonium ion reactions of 2-norbornyl.[7] Such hydride shifts also result in racemization. These hydride shifts are readily observed under stable ion conditions and discussion will be deferred until that topic is considered (Chapter 13).

D. High Exo:Endo Product Ratios

The solvolysis proceeds with almost exclusive formation of *exo* products, both from the *exo* and *endo* isomers. This is attributed to the protection of the *endo* side by the σ-bridge (3).

$$(3)$$

[15]S. Winstein, E. Clippinger, R. Howe, and E. Vogelfanger, *J. Amer. Chem. Soc.*, **87**, 376 (1965).

The fact that *exo* substitution occurs even in the 7,7-dimethyl derivatives was considered to provide major support for the concept (4).[3,16]
Winstein's original position was not as dogmatic in favor of the nonclassical

$$\tag{4}$$

structure as it was later interpreted. Thus, in the original publication[1] he
stated: "... racemization alone could be due to a dynamic equilibrium
between two one-sided cationic species VII and X. Further qualifications
regarding these species and their reactions would be necessary to account for

VII   X

... the essentially exclusive formation of *exo* product and the enhanced
solvolysis rate of the *exo-p*-bromobenzenesulfonate."

6.4. Transition State or Intermediate

It has long been recognized that the rearrangements in bicyclic systems
of the norbornyl type are extraordinarily facile[17] and involve cationic
intermediates.[18] If σ-bridged species are not involved, the barrier for such
interconversion must be very low. If the rate of interconversion of such ions
or ion pairs is to be fast relative to the rate of their capture by solvent, the
barrier to interconversion must indeed be low, not greater than a few
kcal mol^{-1}.

In essence, the nonclassical proposal was that those electronic effects
which served to lower the barrier for the interconversion of two cations
(Figure 6.2) could further stabilize this species so that it produced a
symmetrical intermediate sufficiently stable that there was no longer any
need to consider the original equilibrating unsymmetrical cations.

[16] A. Colter, E. C. Friedrich, N. J. Holness, and S. Winstein, *J. Amer. Chem. Soc.*, **87**, 378 (1965).
[17] G. Wagner and W. Buckner, *Ber.*, **32**, 2302 (1899).
[18] H. Meerwein and K. Van Emster, *Ber.*, **55**, 2500 (1922).

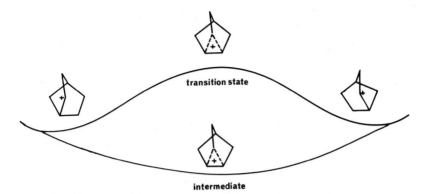

intermediate

Figure 6.2. Proposed transformation of the relatively stable transition state responsible for the rapid equilibration of 2-norbornyl cations into a stabilized symmetrical intermediate sufficiently stable as to make unnecessary further consideration of the unsymmetrical 2-norbornyl cations as intermediates.

This is a fascinating hypothesis. It possesses the advantages of simplicity in replacing two or more equilibrating classical cations by a single σ-bridged species. However, chemistry is still an experimental science—no matter how reasonable and attractive such a proposal may appear, it is still necessary to subject such proposals to experimental test.

6.5. Equilibrating Cations

Some time ago I pointed out a major anomaly in carbonium ion chemistry.[19] Systematic lowering of the potential barrier separating two symmetrical cations (5) would be expected to result in three distinct classes

$$
\underset{+}{>}\overset{R}{\underset{|}{C}}-C< \;\rightleftharpoons\; >C-\overset{R}{\underset{|}{\underset{+}{C}}}< \tag{5}
$$

of such cations: (a) essentially static classical cations, which can be formed and converted into products without significant equilibration; (b) equilibrating cations which undergo rapid equilibration in the time interval between formation and conversion into products; and (c) bridged species where the potential barrier has disappeared so that resonance now occurs involving the two structures. At the time that paper was published (1965), practically all systems examined had been assigned to the first and third of these classes, with the intermediate class being almost unpopulated. It was puzzling why

[19]H. C. Brown, K. J. Morgan, and F. J. Chloupek, *J. Amer. Chem. Soc.*, **87**, 2137 (1965).

there was this apparent discontinuity in the potential barriers separating such pairs of symmetrical cations (5).

Since that time the situation has been greatly changed. Numerous equilibrating cations have been identified in solvolysis, such as 3-phenyl-2,3-dimethyl-2-butyl,[20] 1,2-di-p-anisylnorbornyl,[21] 1,2-dimethyl-norbornyl,[22] etc.

In recent years the pioneering work of Olah, Saunders, Brouwer, and Hogeveen has made possible the direct spectroscopic observation of many cations (Chapter 13). One remarkable development from such studies has been the conclusion that many cations, such as 2,3,3-trimethyl-2-butyl, which can be captured in solvolysis without equilibration,[20] undergo very rapid equilibration under stable ion conditions. Such equilibration often cannot be frozen out even at temperatures as low as −150°C. Indeed, we now have many more ions under stable ion conditions which have been assigned structures as equilibrating classical than have been assigned structures as static classical or σ-bridged.

Let us now consider the argument that the rate of equilibration in the 2-norbornyl cation required to achieve the observed optical behavior is simply too fast to be a reasonable alternative to the σ-bridged interpretation.

The solvolysis of optically active 2-norbornyl brosylate produces acetate which is ≥99.95% racemic.[1] Consequently, the rate of equilibration would need to exceed the rate of collapse of the 2-norbornyl cation by a factor greater than 2000.[23] The rate of ion or ion-pair collapse in aqueous acetone is faster than the 6,2-hydrogen shift. The rate for the latter process at 25°C in nmr solvents[24] was estimated as $10^9 \, \text{sec}^{-1}$ at 25°C. On the assumption that this rate will be the same in acetic acid, the rate of equilibration has been estimated to be in the neighborhood of ≥2 × $10^{12} \, \text{sec}^{-1}$.[2,23] Bartlett[2] concluded that a rate constant of this order of magnitude "happens to be about the rate of passage of a transition state over the barrier in rate theory." In his view, "any reaction occurring so fast has zero activation energy and the migrating carbon would be at just as low an energy when midway in its migration as at the beginning or end." He concluded that "no such picture of decaying molecular structure results if we attribute the racemization to the direct formation of the bridged ion with planar symmetry, and no absurdly high rate constants are required."

[20]H. C. Brown and C. J. Kim, *J. Amer. Chem. Soc.*, **90**, 2082 (1968).

[21]P. v. R. Schleyer, D. C. Kleinfelter, and H. G. Richey, Jr., *J. Amer. Chem. Soc.*, **85**, 479 (1963).

[22]H. Goering and K. Humski, *J. Amer. Chem. Soc.*, **90**, 6213 (1968).

[23]S. Winstein, *J. Amer. Chem. Soc.*, **87**, 381 (1965).

[24]M. Saunders, P. v. R. Schleyer, and G. A. Olah, *J. Amer. Chem. Soc.*, **86**, 5680 (1964).

Recently, Fong has applied the principles of the quantum mechanical theory of relaxation to elucidate the general properties of intramolecular rearrangement processes in carbonium ions.[25] As a result of his analysis, he concludes that a rate of equilibration of $10^{12} \sec^{-1}$ lies well within the expectation of his analysis. Consequently, he concludes that "when the barrier to intramolecular rearrangement is low (i.e., $E_a \leq h\omega$), as indeed appears to be the case in the 2-norbornyl cations, we envision a most elegant mode of quantum mechanical shuttling between two 'classical' wells, far from Bartlett's vision of a 'decaying molecular structure' in which 'C-6 flops about randomly' as if 'in an untidy box'."

Consequently, even a rate as high as $10^{12} \sec^{-1}$ is not incompatible with the Fong treatment. However, there are reasons for believing that this estimate is on the high side. First, the rate of the 6:2 hydrogen shift at 25°C in nmr solvents has now been redetermined to be $10^8 \sec^{-1}$,[26] a factor of 10 lower than the earlier estimate.[24] Second, as discussed later (Chapter 13), there are reasons to expect that the rate of equilibration of cations may be considerably slower in the usual solvolytic media than in media used for nmr observations. Consequently, the original estimate of $\geq 10^{12} \sec^{-1}$ should be lowered to $\geq 10^{11} \sec^{-1}$, and may be lowered considerably further by the solvation factor.

An additional factor to consider is the possibility that solvolysis may produce not the free carbonium ion, but a relatively tight ion pair.[27] The reaction of such an ion pair with solvent may be considerably slower than the rate achieved with free carbonium ions.

From these considerations, we concluded that we could not resolve the nonclassical ion problem from theoretical considerations of the presumed rate of equilibration that would be required to account for the observed degree of racemization.[2,23] But what about the other unique characteristics, exceptionally high *exo* rates, exceptionally high *exo*: *endo* rate ratios, exceptionally high *exo*: *endo* product ratios? Do these not require the σ-bridged formulation?

6.6. Are Exo Rates Unusually Fast?

As discussed previously (Chapter 3), the enhanced rate of solvolysis of camphene hydrochloride (**10**), as compared to *tert*-butyl chloride (1), was considered to be not compatible with relief of steric strain.[13] Consequently,

[25]F. K. Fong, *J. Amer. Chem. Soc.*, **96**, 7638 (1974).
[26]G. A. Olah, A. M. White, J. R. De Member, A. Commeyras, and C. Y. Lui, *J. Amer. Chem. Soc.*, **92**, 4627 (1970).
[27]R. A. Sneen, *Accounts Chem. Res.*, **6**, 46 (1973).

the enhanced rate was attributed to the formation of a stabilized "synarte-tic" ion (**11**).

However, in such comparisons it is essential to select appropriate models. Examination of the related methyl substituted 1-methylcyclopentyl chlorides revealed rates that are comparable to those exhibited by cam-phene hydrochloride and the parent system, 2-methyl-*exo*-norbornyl chloride (6).[14]

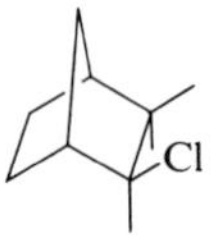

13,600

2380

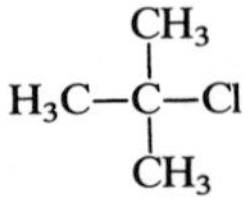

$$H_3C-\underset{\underset{CH_3}{|}}{\overset{\overset{CH_3}{|}}{C}}-Cl \qquad\qquad (6)$$

1.00

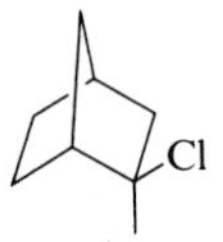

355

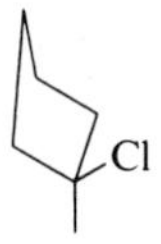

66

Quite clearly, the results are in accord with the postulated effect of increasing steric strain in enhancing the rates of solvolysis of highly branched tertiary chlorides (Section 2.2). Similar results have been realized in other comparisons.[28] Consequently, it does not appear that such rates provide an unambiguous basis on which to base proposals for σ-bridging (Chapter 3).

6.7. Do the High Exo:Endo Rate Ratios Require σ-Bridged Cations?

The high *exo*: *endo* rate ratio exhibited by 2-norbornyl is a remarkable phenomenon, perhaps the least ambiguous characteristic of the 2-norbornyl system on which to base a decision. The precise problem of the 2-norbornyl solvolysis is defined by the energetics of the system as defined by the energy diagram introduced by Goering and Schewene (Figure 6.3).[29]

[28]H. C. Brown, F. J. Chloupek, and M.-H. Rei, *J. Amer. Chem. Soc.*, **86**, 1246, 1247, 1248 (1964).
[29]H. L. Goering and C. B. Schewene, *J. Amer. Chem. Soc.*, **87**, 3516 (1965).

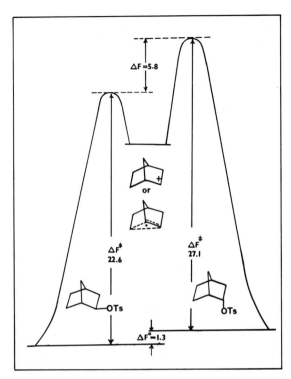

Figure 6.3. Free-energy diagram for the acetolysis of *exo*- and *endo*-norbornyl tosylates.

The rate of acetolysis of *exo*-norbornyl brosylate is 350 times that for the corresponding *endo* derivative.[1] If the data are corrected for internal return in the *exo* isomer, then the *exo*:*endo* rate ratio becomes 1600. The relative rate of 1600 corresponds to a difference in the free energy of activation of 4.5 kcal mol⁻¹. The strain in *endo*-norbornyl arenesulfonates is estimated to be 1.3 kcal mol⁻¹.[30] This leads to a difference in the energies of the two transition states of 5.8 kcal mol⁻¹ (Figure 6.3). Following the transition states, both isomeric brosylates or tosylates yield a common intermediate. The problem we are facing is that of defining just what factor or factors are responsible for the difference in energies of the two transition states. Is the *exo* transition state stabilized by σ-bridging? Or is it the *endo* transition state that is anomalous—destabilized by steric hindrance to ionization (Chapter 8)?

It appeared that a reasonable solution to this problem would be the synthesis of a 2-norbornyl cation sufficiently stabilized by a group such as 2-*p*-anisyl that σ-participation could not be a factor. Then if the *exo*:*endo* rate

[30] P. v. R. Schleyer, *J. Amer. Chem. Soc.*, **86**, 1854, 1856 (1964).

ratio dropped to one, the σ-bridged interpretation in the parent system would be supported. On the other hand, if the *exo*:*endo* ratio remained high, then the *exo*:*endo* rate ratio must be a steric characteristic of the 2-norbornyl system and not the result of σ-bridging.

6.8. Do the High Exo:Endo Product Ratios Require σ-Bridged Cations?

It is evident from Figure 6.3 that recombination of the ion or ion-pair in the central well with the anion will occur preferentially to produce the *exo* ester as compared to the *endo* ester.

What we would like to know is not the relative rates at which the cationic intermediate reacts with the anion to reform the original esters, but the relative rates at which the intermediate captures the solvent acetic acid to form the two products, *exo*- and *endo*-norbornyl acetate. Fortunately, the data are available. Goering and Schewene[29] measured the rates of racemization, the rates of exchange, and the equilibration of *exo*- and *endo*-norbornyl acetate. The results of their study is shown in the free-energy diagram (Figure 6.4).

We are now in position to consider what happens to the cationic intermediate in reacting with acetic acid solvent to form the two acetates. The energy of activation of the ionization of the *endo* acetate is $4.5\,\text{kcal mol}^{-1}$ higher than for the *exo*. If we correct for the higher ground-state energy of the *endo* isomer, $1.0\,\text{kcal mol}^{-1}$, the difference in the energies of the two transition states becomes $5.5\,\text{kcal mol}^{-1}$.

The two reactants pass over these two transition states to form the same intermediate, the norbornyl cation or ion pair. The principle of microscopic reversibility requires that in the symmetrical system (the subject of this discussion) the norbornyl intermediate, in reacting with the solvent, must pass over the same two transition states that are involved in the solvolysis of the reactants. The authors reported that the product from the solvolysis of 2-norbornyl brosylate consists of 99.98% *exo*- and 0.02% *endo*-norbornyl acetate. This is in reasonable agreement with the distribution predicted for a difference of $5.5\,\text{kcal mol}^{-1}$ in the energies of the two transition states.

It follows that the factor responsible for the difference in energy between the transition states for the *exo* and *endo* isomers must be largely responsible for the difference in the *exo* and *endo* rates of solvolysis. (The difference in ground-state energies also contributes, but this difference is relatively small in the present case compared to the much larger difference in the energies of the two transition states.) It also follows that the factor responsible for the difference in energy between the *exo* and *endo* transition

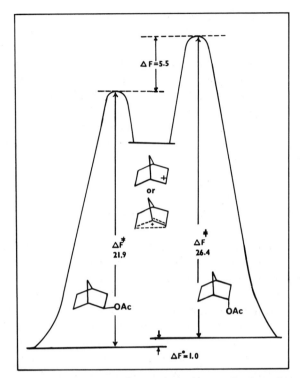

Figure 6.4. Free-energy diagram for the acetolysis of *exo*- and *endo*-norbornyl acetates.

states must be responsible for the stereoselectivity leading to the almost exclusive formation of the *exo* product. Unfortunately, the fact that we recognize this point does not aid us in understanding just what that factor may be.

The remarkable stereoselectivity exhibited in the solvolysis of 2-norbornyl derivatives has been considered to require bridging in the norbornyl cation.[1,23,29] However, the diagram makes it clear that the amount of bridging that may or may not be present in the free ion is not directly involved in the stereoselectivity of the substitution. *It is the amount of bridging in the exo transition state, or whatever the factor responsible for the difference in the two transition states, that will control the distribution of the norbornyl cation between exo and endo product.*

Let us consider what might be the factor or factors responsible for the difference in energy between the *exo* and *endo* transition states for the secondary system. Four possibilities may be pointed out: (a) the *exo* transition state is stabilized by σ-participation, with the *endo* being normal; (b) the *endo* transition state is destabilized by steric strain, with the *exo* transition state being normal; (c) a combination of (a) and (b); or (d) some other factor, such as torsional effects,[31] not now considered by current theory.

Here also it appeared that a reasonable solution to the problem would be the synthesis and study of 2-*p*-anisyl-2-norbornyl. It was clear that σ-bridging could not be significant in such a stable cation.[23] Would such derivatives yield high *exo*:*endo* product ratios? If so, then this must be a steric characteristic of the 2-norbornyl system. If a low *exo*:*endo* product ratio were realized, then the results would support the σ-bridging interpretation in the parent system.

6.9. Theoretical and Empirical Solutions

The Goering–Schewene diagram makes the 2-norbornyl problem crystal clear. Is the *endo* transition state normal, with the *exo* transition state stabilized by $5.8 \, \text{kcal mol}^{-1}$ (Figure 6.3)? Or is the *exo* transition state normal, with the *endo* transition state destabilized by $5.8 \, \text{kcal mol}^{-1}$? Consequently, a possible solution to the problem would be by a theoretical calculation of the energies of the two transition states in Figure 6.3. Unfortunately, this appears to be beyond us at the present time.

Since the transition states of solvolytic processes and the intermediates are closely related, according to the Hammond postulate,[32] perhaps we could approach the problem by calculating the relative energies of the unsolvated classical and σ-bridged 2-norbornyl cations. A number of such calculations have been reported and it is interesting to observe the changes with the years.

Extended Hückel theory was applied to 2-norbornyl. It indicated the most stable structure to be the σ-bridged structure,[33] with minimum energy when C6 is sp^2 hybridized.[34] The calculation indicated a stabilization of $51 \, \text{kcal mol}^{-1}$ for the bridged structure relative to the classical unbridged structure.

Klopman, using the LCAOSCF method, found the nonclassical structure to be favored over the classical structure by some $40 \, \text{kcal mol}^{-1}$.[35] His calculation favored an edge-protonated structure (**4**) over the face protonated species (**5**).

[31]P. v. R. Schleyer, *J. Amer. Chem. Soc.*, **89**, 699 (1967).
[32]G. S. Hammond, *J. Amer. Chem. Soc.*, **77**, 334 (1955).
[33]R. Hoffman, *J. Chem. Phys.*, **39**, 1397 (1963); *ibid.*, **40**, 2480 (1964).
[34]W. S. Trahanovsky, *J. Org. Chem.*, **30**, 1666 (1965).
[35]G. Klopman, *J. Amer. Chem. Soc.*, **91**, 89 (1969).

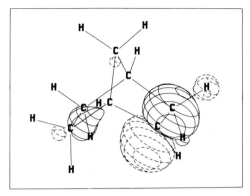

Figure 6.5. LUMO for the Dewar (MINDO/3) 2-norbornyl cation.

More recently, Goetz and Allen have utilized *ab initio* calculations to conclude that the total energy of the classical structure for the 2-norbornyl cation (**1**) is 4.7 kcal mol^{-1} lower than that of the σ-bridged structure (**3**).[36] However, they also estimated that the "addition of polarization functions will produce a further lowering of the nonclassical relative to the classical structure, most likely making the nonclassical stable."[37]

Finally, M. J. S. Dewar and coworkers have applied MINDO/3 to the problem. They have concluded that the classical structure is more stable than the nonclassical formulation by 2 kcal mol^{-1}. Although they feel that this is "too close to call," they conclude that the difference in stabilities must be relatively small, so that one cannot account for the *exo*: *endo* rate ratio in terms of nonclassical contributions.[38] The LUMO[39] of the Dewar structure is shown in Figure 6.5.

At one time the Foote–Schleyer correlation appeared to offer hope of resolving the question.[30] This empirical correlation provides a means of calculating the rates of acetolysis of secondary tosylates. Clearly, if the treatment could be relied upon to reveal that the *endo* isomer is normal, whereas the *exo* isomer is 350 times more reactive than the calculated value, the problem could be considered solved. Indeed, the initial application of the correlation appeared to point to that conclusion.[30]

Study of the correlation indicated that a possible difficulty lay in the treatment of steric hindrance to ionization. In principle, the nonbonded

[36]D. W. Goetz and L. C. Allen, *XXIII International Congress of Pure and Applied Chemistry*, Vol. 1, Butterworth, London, 1971, p. 51.

[37]D. W. Goetz and L. C. Allen, *J. Amer. Chem. Soc.*, submitted.

[38]M. J. S. Dewar, R. C. Haddon, A. Komornicki, and H. Rzepa, *J. Amer. Chem. Soc.*, **99**, 377 (1977).

[39]W. L. Jorgensen and L. Salem, *The Organic Chemist's Book of Orbitals*, Academic Press, New York, 1973.

strain term $(GS_{strain} - TS_{strain})$ in the correlation should be capable of handling steric hindrance to ionization $(GS_{strain} < TS_{strain})$ as well as steric assistance to ionization $(GS_{strain} > TS_{strain})$. However, in practice the expediency had been adopted of assuming $TS_{strain} \approx 0$ for leaving groups.

Accordingly, a joint program was undertaken with Schleyer to test the ability of the correlation to predict the rates of certain *endo*-norbornyl derivatives (7) where the steric hindrance to ionization might be expected to be relatively large (as discussed in Chapter 8).

$$\text{(7)}$$

14   **15**   **16**

The discrepancies between the predicted and observed rates were 8000 for **14**, 10,000 for **15**, and 100,000 for **16**. (For a more detailed discussion see Section 8.6.) It was concluded that the correlation in its present state of development could not be relied upon to provide a definitive answer to the nonclassical ion problem.[40]

6.10. Conclusion

Since neither theoretical nor empirical approaches appeared capable of resolving the problem, we decided to undertake the synthesis of highly stabilized 2-*p*-anisyl-2-norbornyl and 2-*p*-anisyl-2-camphenilyl derivatives for examination of the *exo*:*endo* rate and product ratios. Such stabilized derivatives cannot involve σ-bridging.[23] Would these derivatives exhibit low *exo*:*endo* rate and product ratios, supporting the σ-bridged interpretation of 2-norbornyl, or would they exhibit high *exo*:*endo* rate and product ratios?

[40]H. C. Brown, I. Rothberg, P. v. R. Schleyer, M. M. Donaldson, and J. J. Harper, *Proc. Nat. Acad. Sci.*, **56**, 1653 (1966).

Comments

The high *exo:endo* ratios exhibited by *secondary* 2-norbornyl derivatives should be considered in the context of the lower epimeric rate ratios typically found in other *secondary* ring systems where neighboring groups are absent. Cyclohexyl *axial/equatorial* ratios, in a great variety of molecules containing 6-membered rings, are usually in the range 2–4, some 10^2–10^3 smaller than the norbornyl value. Epimeric ratios in cyclopentanes are even less, e.g., in the 9-tetrahydro-*endo*-dicyclopentadienyl series (8).[40]

RR (HOAc, 25°C):	1.0	1.6

Even at the sterically more critical 8-position, a low epimeric ratio is found (9).[40] Perhaps solvent assistance is important in **16**, but not in **17**, and this reduces the ratio. It is also possible that such ratios are inherently low in nonparticipating systems.

	16	**17**
RR (HOAc, 25°C):	1.0	5.7

I have challenged Professor Brown to provide examples of high epimeric ratios in *secondary* systems when anchimeric assistance is clearly absent (no neighboring group, products not rearranged). As yet, this challenge has not been met.

Goetz and Allen[37] have carried out further optimization of the 2-norbornyl cation at both the STO-3G and 4-31G levels with the result that bridged (**3**) and classical (**1**) structures are now almost exactly equal in energy. Since larger basic sets and the inclusion of correlation (which is not yet possible with a molecule as large as $C_7H_{11}^+$) are known from experience with smaller ions to favor bridged over classical structures, they conclude that the 2-norbornyl cation is bridged (**3**).

The MINDO/3 calculations of Dewar[38] have inconsistencies which reduce confidence in their reliability. The calculated heat of formation for the parent hydrocarbon, norbornane, is in error by 20.8 kcal/mole![41] This discrepancy may be important. In my view, the norbornyl cation bridges because of strain relief. The longer nonclassical bonds permit relaxation of some of the unfavorable bond angles.[42] The 1,7-delocalized bridged ion **18**, for which no experimental evidence exists, is actually found by MINDO/3 to be 3.5 kcal/mole more stable than **1**.[38] Finally, according to MINDO/3, the 7-norbornyl cation (**19**), a species legendary in its inertness,[30] is found erroneously to be 3.1 kcal/mole *more* stable than the 2-cation.[43]

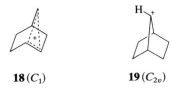

$$18\,(C_1) \qquad\qquad 19\,(C_{2v})$$

Such theoretical calculations, of course, refer to the isolated state (gas phase). Whether or not solvation preferentially stabilizes classical or bridged structures has yet to be established.[44] Although I feel this is unlikely in the case of 2-norbornyl, one cation form may be favored in the gas phase and another in solution. This possibility is indicated by recent calculations.[43]

While it is true that the Foote–Schleyer correlation cannot be employed with confidence in situations where the leaving group is crowded (e.g., **14–16**), this criticism does not pertain to 2-*exo*-norbornyl tosylate, where the tosylate is in a strain-free position. Although high confidence should not be placed in Foote–Schleyer predictions, I feel that our subsequent work with Brown[40] does not provide grounds to doubt the original conclusion[30] that 2-*exo*-norbornyl tosylate reacts more rapidly than expected.

[*Comment added in proof* (HCB)] Jorgensen has recently concluded on the basis of MINDO/3 calculations that solvation stabilizes classical cations to a greater extent than nonclassical cations.[44]

[41]R. C. Bingham, M. J. S. Dewar, and D. H. Lo, *J. Amer. Chem. Soc.*, **97**, 1294 (1975).
[42]L. Radom, J. A. Pople, V. Buss, and P. v. R. Schleyer, *J. Amer. Chem. Soc.*, **94**, 311 (1972).
[43]W. L. Jorgensen, private communication.
[44]W. L. Jorgensen, *J. Amer. Chem. Soc.*, **99**, 280 (1977).

7

Stabilized 2-Norbornyl Cations

7.1. Introduction

The position that the importance of neighboring group participation should diminish as an incipient cationic center is stabilized by substitution[1] appears to be generally accepted.[2] Consequently, we undertook to synthesize representative 2-norbornyl systems containing a stabilizing *p*-anisyl group at the 2-position in order to establish its effect on the *exo* : *endo* rate and product ratios.

7.2. The Gassman–Fentiman Approach

The validity of this position has been beautifully confirmed by Gassman and Fentiman in their study of the effect of such stabilizing groups[3] on the enormous π-participation present in *anti*-7-norbornenyl derivatives[4] (Chapter 4).

The rate enhancement of 10^{11} observed in the parent secondary derivatives decreases with the introduction of stabilizing groups at the 7-position (2) and effectively vanishes with the *p*-anisyl group (1).

[1]S. Winstein, B. K. Morse, E. Grunwald, K. C. Schreiber, and J. Corse, *J. Amer. Chem. Soc.*, **74**, 1113 (1952).

[2]P. D. Bartlett, *Nonclassical Ions*, Benjamin, New York, 1965.

[3]P. G. Gassman and A. F. Fentiman, *J. Amer. Chem. Soc.*, **90**, 2691 (1968).

[4]S. Winstein, M. Shatavsky, C. Norton, and R. B. Woodward, *J. Amer. Chem. Soc.*, **77**, 4183 (1955).

	1	**2**	(1)
Z = 7-H	1.00	10^{11}	
3,5-$(CF_3)_2$	1.00	255,000	
p-CF_3	1.00	34,000	
p-H	1.00	41.5	
p-CH_3O	1.00	3.4	

If *p*-anisyl can cause the truly enormous π-participation ($\times 10^{11}$) observed in *anti*-7-norbornenyl to vanish, it should also cause the much smaller σ-participation ($\times 350$) proposed for *exo*-norbornyl to vanish.

7.3. Exo:Endo Rate Ratios in 2-p-Anisyl-2-norbornyl

Addition of *p*-anisylmagnesium bromide to 2-norbornanone (**3**) yielded 2-*p*-anisyl-*endo*-norbornanol (**4**). Treatment with hydrogen chloride gave the *exo*-chloride (**5**). Hydrolysis then yielded the 2-*p*-anisyl-*exo*-norbornanol (**6**). The transformations (2) are typical of the usual reactions of norbornyl derivatives.[5,6]

[5]H. C. Brown and K. Takeuchi, *J. Amer. Chem. Soc.*, **88**, 5336 (1966).
[6]H. C. Brown and K. Takeuchi, *J. Amer. Chem. Soc.*, **90**, 2691 (1968).

Solvolysis of the benzoates (the *p*-nitrobenzoate of the *exo* isomer (**7**) is unstable) and *p*-nitrobenzoates in 80% acetone revealed the *exo* : *endo* rate ratio (**7**:**8**) to be 284 (3). This compares to a value of 350 for the parent 2-norbornyl brosylates[7] and 280 for the corresponding tosylates.[8]

(3)

RR (25°C):	284	1.00

Consequently, we do not observe a significant decrease in the *exo* : *endo* rate ratio, in spite of the high stability of the cationic center, which should make σ-bridging insignificant.

A pmr study of the 2-phenyl-2-norbornyl cation has revealed no evidence for charge delocalization from the 2- to the 1- and 6-positions.[9] If the 2-phenyl-2-norbornyl cation is classical, there surely cannot be any argument about the more stable 2-*p*-anisyl derivative.

Clearly, the results do not support the interpretation that high *exo* : *endo* rate ratios in the 2-norbornyl system must be the result of σ-participation, even in tertiary derivatives.[10]

7.4. Exo : Endo Rate Ratios in 2-p-Anisyl-2-camphenilyl

The 2-*p*-anisyl-2-camphenilols (*exo* and *endo*) dissolve in acid to give a particularly stable cation.[11] Accordingly, we synthesized the *exo*-benzoate and the *endo*-*p*-nitrobenzoate and determined the *exo* : *endo* rate ratio in 80% acetone. Here the *exo* : *endo* rate ratio (**9**:**10**) increases to 44,000 (4).[12]

[7]S. Winstein and D. Trifan, *J. Amer. Chem. Soc.*, **74**, 1147, 1154 (1952).

[8]P. v. R. Schleyer, M. M. Donaldson, and W. E. Watts, *J. Amer. Chem. Soc.*, **87**, 375 (1965).

[9]D. G. Farnum and G. Mehta, *J. Amer. Chem. Soc.*, **91**, 3256 (1969). Some nmr evidence for charge delocalization is observed in derivatives with more electron-demanding substituents; see D. G. Farnum and A. D. Wolf, *J. Amer. Chem. Soc.*, **96**, 5166 (1974).

[10]S. Winstein, *J. Amer. Chem. Soc.*, **87**, 381 (1965).

[11]P. D. Bartlett, E. R. Webster, C. E. Dills, and H. G. Richey, *Ann.*, **623**, 217 (1959).

[12]H. C. Brown and K. Takeuchi, *J. Amer. Chem. Soc.*, **90**, 5268 (1968).

RR (25°C):   44,000   1.00

An examination of the individual rate constants (Table 7.1) reveals that the rates of the *exo* isomers (**7** and **9**) are quite similar. The high *exo* : *endo* rate ratio results largely from a considerably decreased rate for the *endo* isomer (**10**) (5). A comparison of the rates of acetolysis of *endo*-camphenilyl

RR (25°C):
                  1.00   1/147

brosylate (**12**) with *endo*-norbornyl brosylate (**11**) also reveals a decrease for the camphenilyl derivative (**6**),[13] albeit somewhat smaller.

RR (25°C):   1.00   1/10

These decreases are consistent with the proposal of decreased rates of solvolysis for the *endo* isomers because of the U-shaped cavity of the *endo* face (Chapter 8). The *endo*-3-methyl substituent should contribute to trapping the departing anion in the *endo* cavity.

7.5. Exo:Endo Product Ratios

The solvolysis of 2-*p*-anisyl-*exo*-norbornyl chloride (**5**) yields 2-*p*-anisyl-*exo*-norbornanol (**6**).[5] Unfortunately, the product is highly unstable and difficult to analyze quantitatively. Solvolysis of the chloride (**5**) in the presence of sodium borohydride (**7**)[14] gives a more stable product, the

[13] A. Colter, E. C. Friedrich, N. J. Holness, and S. Winstein, *J. Amer. Chem. Soc.*, **87**, 378 (1965).
[14] H. M. Bell and H. C. Brown, *J. Amer. Chem. Soc.*, **88**, 1473 (1966).

TABLE 7.1. Rates of Solvolysis of Representative 2-p-Anisyl Derivatives

p-Nitrobenzoate	Isomer	$10^6 k_1$, sec^{-1} 25°	Exo:endo rate ratio	$\Delta H^{\ddagger}$, kcal mol^{-1}	$\Delta S^{\ddagger}$, eu
2-p-Anisyl-2-norbornyl	exo	11,400[a]	284		
	endo	40.2			
2-p-Anisyl-2-camphenilyl	exo	12,100[a]	44,000	26.0	−1.5
	endo	0.273[b]			
2-p-Anisyl-2-norbornenyl	exo	2,520[a]	312	22.8	−5.6
	endo	8.08			
2-p-Anisyl-5-methyl-2-norbornenyl	exo	3,000[a]	354		
	endo	8.48			
2-p-Anisyl-2-benzo-norbornenyl	exo	1,080	3,300	24.5	−6.0
	endo	0.328[b]			
2-p-Anisyl-6-methoxy-2-benzonorbornenyl	exo	2,190[a]	7,000	25.1	−4.0
	endo	0.311			

[a] Estimated by multiplying the rate of the benzoate by the factor 20.8.[6]
[b] Calculated from data at other temperatures.

hydrogen derivative (**13**). Analysis by pmr reveals attack by the borohydride from the *exo* direction, ≥98%.[6] Consequently, even this highly stabilized cation exhibits predominant substitution from the *exo* side.

The presence of the *gem*-dimethyl substituents in 2-p-anisyl-2-camphenilyl greatly simplifies the analysis by pmr and permits a precise determination. Solvolysis of either isomer in 80% acetone (8) proceeds to give ≥99.5% of the *exo*-alcohol (**14**).[15]

[15] K. Takeuchi and H. C. Brown, *J. Amer. Chem. Soc.*, **90**, 5270 (1968).

Even in the much more hindered 1,7,7-trimethyl-2-*p*-anisylnorbornyl cation (**16**), produced in the solvolysis of 1-*p*-anisylcamphenehydrochloride (**15**), there is a preference for *exo* attack (9),[16] although far less so than in the simpler cases just discussed.

$$(9)$$

It was one of the tenets of the nonclassical ion theory that 7,7-dimethyl substituents would exclude *exo* substitution in the absence of σ-bridging.[17] Clearly that cannot be the case here.

7.6. Goering–Schewene Diagrams

Consequently, even these highly stabilized 2-norbornyl derivatives exhibit the high *exo* : *endo* rate and product ratios previously considered to represent a major argument favoring the σ-bridged formulation. If we accept both the theoretical and the experimental evidence that highly stabilized cations, such as 2-*p*-anisyl-2-norbornyl, cannot involve σ-bridging, it follows that high *exo* : *endo* rate and product ratios are not evidence for such σ-bridging. More logically, there must be some characteristic feature of the norbornyl structure, other than σ-bridging, that is responsible for the observed high *exo* : *endo* rate and product ratios.

It is of interest to construct the Goering–Schewene diagrams for the 2-*p*-anisyl-2-norbornyl and the 2-*p*-anisyl-2-camphenilyl systems for comparison with the diagram for the parent system (Figure 6.3).

The free energy of activation for the solvolysis of 2-*p*-anisyl-*exo*-norbornyl *p*-nitrobenzoate in 80% aqueous acetone is

[16]H. C. Brown and H. M. Bell, *J. Amer. Chem. Soc.*, **86**, 5006, 5007 (1964).

[17]J. A. Berson, *Molecular Rearrangements*, Vol. 1, P. de Mayo, Ed., Interscience, New York, 1963, Chapter 3.

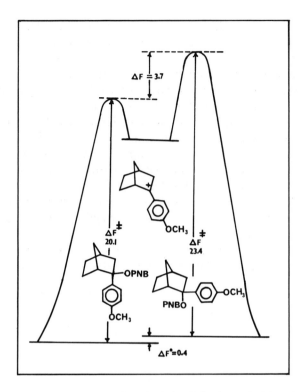

Figure 7.1. Free-energy diagram for the solvolysis of the 2-*p*-anisyl-2-norbornyl *p*-nitro-benzoates in 80% acetone at 25°C.

20.1 kcal mol^{-1}.[6] The corresponding value for the *endo* isomer is 23.4 kcal mol^{-1}. Both isomers yield the same intermediate. Equilibration of the two epimeric alcohols in the case of the phenyl derivatives established that they are of considerable stabilities, with the ground state of the *endo* derivative being higher in energy by a relatively small factor of 0.4 kcal mol^{-1} [18] Finally, the available evidence indicates that there is no significant difference in the steric requirements of the acyloxy and hydroxyl groups in the norbornyl system.[18]

With these approximations in mind, it is possible to construct a free energy diagram[19] for the solvolysis of the *exo* : *endo* pair (Figure 7.1). The diagram reveals a difference in energy of the two transition states of 3.7 kcal mol^{-1}.

It is clear that the cation once formed will react with the anion (or the solvent) to give the two epimeric derivatives in a ratio determined by the

[18]M.-H. Rei and H. C. Brown, *J. Amer. Chem. Soc.*, **88**, 5335 (1966).
[19]H. L. Goering and C. B. Schewene, *J. Amer. Chem. Soc.*, **87**, 3516 (1965).

difference in energy in the respective transition states. The reaction will evidently proceed in the case of both isomers to give the *exo* isomer predominantly.

The essential problem is that of accounting for the major difference in the energies of the two transition states. Is the *endo* transition state normal, with the *exo* transition state stabilized by some factor, or is the *exo* transition state normal with the *endo* transition state destabilized by some factor?

In the case of the parent system, 2-norbornyl itself (Figure 6.1), it has been customary to argue that the *exo* transition state must be stabilized by σ-bridging over the corresponding *endo* transition state.[19] However, this explanation cannot be utilized for the 2-*p*-anisyl derivative. There appears to be considerable experimental evidence and general agreement that a cationic center highly stabilized by a 2-*p*-anisyl group cannot engage in such σ-bridging.

This leaves us with the possibility that the *exo* transition state must be considered normal, with the *endo* transition state destabilized by some

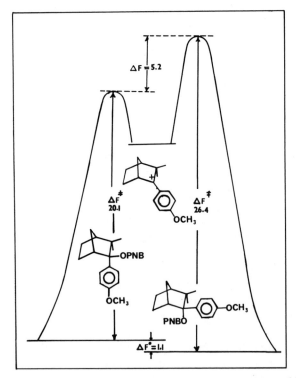

Figure 7.2. Free-energy diagram for the solvolysis of the 2-*p*-anisylcamphenilyl *p*-nitro-benzoates in 80% acetone at 25°C.

factor. Steric hindrance to ionization has been suggested as that factor and will be discussed later (Chapter 8).

In the case of 2-*p*-anisyl-*exo*-camphenilyl *p*-nitrobenzoate, the free energy of activation is the same, 20.1 kcal mol^{-1}, as for the corresponding 2-norbornyl derivative.[2] However, the free energy of activation for the *endo* isomer is 26.4 kcal mol^{-1}, considerably larger than the value for the corresponding 2-norbornyl derivative. Consequently, as pointed out earlier, the major increase in the *exo* : *endo* rate ratio observed for the 2-*p*-anisyl-2-camphenilyl derivatives (44,000), as compared with the 2-*p*-anisylnorbornyl derivatives (284), comes about primarily because of a major decrease in the rate of solvolysis of the *endo* isomer of the camphenilyl system. This decrease is attributed to additional hindrance to the ionization of the *endo*-leaving group provided by the *endo*-3-methyl substituent reinforced by a tilted conformation of the 2-aryl group induced by the *gem*-dimethyl substituents.

Construction of the free energy diagram (Figure 7.2) reveals a difference in the energies of the two transition states of 5.2 kcal mol^{-1}, considerably larger than in the related 2-*p*-anisyl-2-norbornyl system (Figure 7.1). It follows that the partition of the cationic intermediate between *exo* and *endo* product should be more stereoselective for *exo* in the case of the camphenilyl derivative than in the case of the corresponding norbornyl derivative (Figures 7.1 and 7.2).

7.7 Other 2-p-Anisyl Systems

The acetolysis of 2-norbornenyl brosylates (10) exhibits an *exo* : *endo* rate ratio of 7000.[20] The solvolysis product is predominantly 3-nortricyclyl

	(10)
17 H	**18** OBs

RR (25°C):   7000   1.00

acetate (11).[20] Consequently, it is reasonable that the reaction proceeds with π-participation, and the observed *exo* : *endo* rate ratio was attributed to such participation.[20]

$$ \text{(11)} $$

[20]S. Winstein, H. Walborsky, and K. C. Schreiber, *J. Amer. Chem. Soc.*, **72**, 5795 (1950).

An alternative method of examining the data is to compare the unsaturated derivatives to the corresponding saturated compounds (12). The observed decrease in rate of 44 is attributed to the inductive effect of the double bond.

$$\tag{12}$$

RR (25°C):   1.00   1.0/44.0

In the *exo* derivative, this factor decreases to 1.96 (13).

$$\tag{13}$$

RR (25°C):   1.00   1.0/1.96

If we accept the conclusion indicated by our studies that σ-participation is not a factor in 2-norbornyl, then π-participation in **17** is only a factor of 22.

π-Participation in 2-norbornenyl is evidently a very small factor compared to the 10^{11} factor observed in *anti*-7-norbornenyl (1). π–Participation should surely vanish with the 2-*p*-anisyl derivatives. Indeed, both *exo* and *endo* isomers solvolyze in 80% acetone to give the open *exo* alcohols (14).

$$\tag{14}$$

Yet the *exo* : *endo* rate ratio does not vanish to 1, but remains high at 312 (15).[21]

$$\tag{15}$$

RR (25°C):   312   1.00

[21]H. C. Brown and E. N. Peters, *J. Amer. Chem. Soc.*, **97**, 7442 (1975).

This *exo* : *endo* factor of 312 cannot be the result of π-participation. Presumably it arises from steric effects (Chapter 8). If we can carry this factor from the tertiary to the secondary derivatives, then the *exo* : *endo* rate ratio of 7000 observed for the secondary derivatives (10) can be dissected into a factor of 312 attributed to the steric factor and one of 22 attributed to π-participation.

Activation of the double bond by a methyl substituent greatly increases the *exo* : *endo* rate ratio (16).[22]

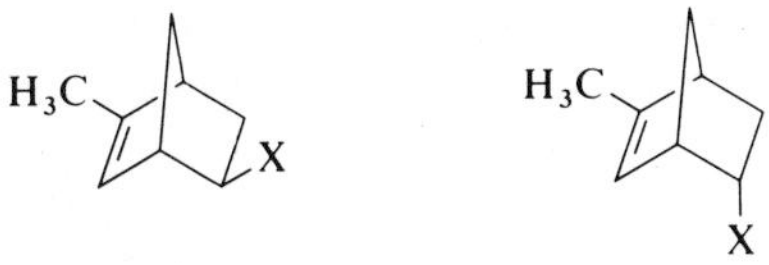

(16)

RR (25°C):   22,000,000   1.00

Yet a 2-*p*-anisyl group also swamps this huge π-participation (17).[23]

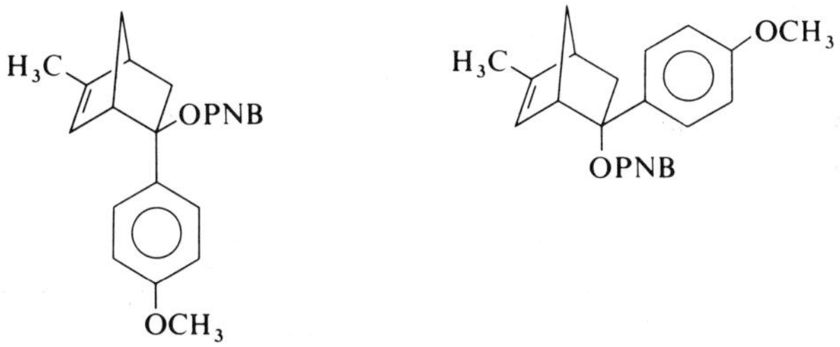

(17)

RR (25°C):   354   1.00

2-Benzonorbornenyl brosylates exhibit a high *exo* : *endo* rate ratio of 15,000[24] (62,000 corrected for internal return[25]) (18). Again the question

(18)

RR (25°C):   15,000   1.00

arises as to how much of this effect should be attributed to aryl participation and how much to steric hindrance to ionization.

[22]Private communication from C. F. Wilcox.
[23]H. C. Brown, E. N. Peters, and M. Ravindranathan, *J. Amer. Chem. Soc.*, **97**, 7449 (1975).
[24]H. C. Brown and G. L. Tritle, *J. Amer. Chem. Soc.*, **90**, 2689 (1968).
[25]J. P. Dirlam, A. Diaz, S. Winstein, W. P. Giddings, and G. C. Hanson, *Tetrahedron Lett.*, 3133 (1969).

Introduction of a methoxy group into the homopara position greatly increases the *exo* : *endo* rate ratio (19).[24,26] Here also the introduction of a 2-

			(19)
RR (25°C):	900,000	1.00	

p-anisyl group does not wipe out the *exo* : *endo* rate ratio, but reduces it to a factor in the neighborhood of 3300[27] for **(19)** and 7000 for **(20)** **(20)**.[28]

			(20)
Exo : *endo*	3300	7000	

These results are summarized in Table 7.1.

Every indication is that the introduction of a highly stabilizing group at a carbonium ion center swamps the ability of that center to interact with a neighboring group. Yet we observe high *exo* : *endo* rate ratios for these 2-*p*-anisyl derivatives: 284 for 2-*p*-anisyl-2-norbornyl, 312 for 2-*p*-anisyl-2-norbornenyl, 354 for 2-*p*-anisyl-5-methyl-2-norbornenyl, 3300 for 2-*p*-anisyl-2-benzonorbornenyl, and 7000 for 2-*p*-anisyl-6-methoxyl-2-benzonorbornenyl. These high *exo* : *endo* rate ratios cannot be the result of significant σ- or π-participation. It has been suggested that these ratios must arise from steric hindrance to ionization attributed to the unique rigid U-shaped structure of the norbornyl system (Chapter 8).

7.8. Other Stabilized Systems.

Another way of stabilizing the 2-norbornyl cation is by the introduction of a methoxy substituent in the 2-position (21).[29] Traylor and Perrin argued

[26]D. V. Braddon, G. A. Wiley, J. Dirlam, and S. Winstein, *J. Amer. Chem. Soc.*, **90**, 1901 (1968).
[27]H. C. Brown, S. Ikegami, K.-T. Liu, and G. L. Tritle, *J. Amer. Chem. Soc.*, **98**, 2531 (1976).
[28]H. C. Brown and K.-T. Liu, *J. Amer. Chem. Soc.*, **91**, 5909 (1969).
[29]T. G. Traylor and C. L. Perrin, *J. Amer. Chem. Soc.*, **88**, 4934 (1966).

$$(21)$$

that such a stabilized cation could not be σ-bridged and could be used to establish *exo* : *endo* rate and product ratios for a truly classical norbornyl cation.

They observed that the acid-catalyzed exchange of methanol-d_4 with norcamphor dimethyl ketal proceeded to give the *exo* isomer preferentially **22** (22). The capture of methanol by the 2-methoxynorbornyl ion (**21**) is

$$(22)$$

simply the reverse of acid-catalyzed ionization. Consequently, the ionization rate is equal to the exchange ratio, *exo* : *endo* = 16. The authors attribute the lower rate of ionization of the *endo*-methoxy group to steric hindrance to ionization. They also trapped the ion (**21**) with lithium borohydride (23). They concluded that the small amount of *endo* capture of

$$(23)$$

the nucleophile must be the result of steric hindrance to nucleophile capture. However, they decided that the ratio of approximately 16:1 observed for the ionization must represent the true *exo* : *endo* rate and product ratio for a truly "classical" 2-norbornyl cation, so that higher ratios, such as 284 observed for 2-*p*-anisyl-2-norbornyl (3) and 44,000 observed for 2-*p*-anisyl-2-camphenilyl (4) must arise from σ-bridging even in these highly stabilized systems.

An alternative interpretation of the low *exo* : *endo* ratio in the ketals is presented in the next section.

Another means of stabilizing the positive charge in the 2-position has been suggested.[30,31] It is proposed that the introduction of an *exo*-double

[30]C. F. Wilcox and R. G. Jesaitis, *Tetrahedron Lett.*, 2567 (1967).
[31]F. C. Wilcox and R. G. Jesaitis, *Chem. Commun.*, 1046 (1967).

bond (**24**) or a spirocyclopropane ring (**25**) at the 3-position should so delocalize positive charge from the 2-position as to produce secondary

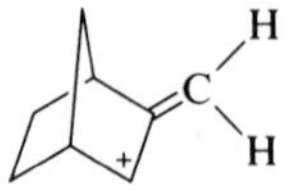

2-norbornyl cations so stable as to eliminate σ-participation. Indeed, low *exo* : *endo* rate ratios are observed in **26–29** (24).[30–32]

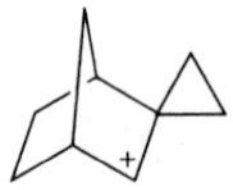

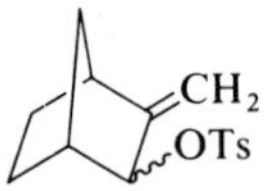

Exo : *endo*	3.9	3.2

(24)

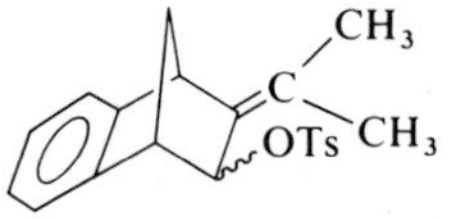

Exo : *endo*	8	12

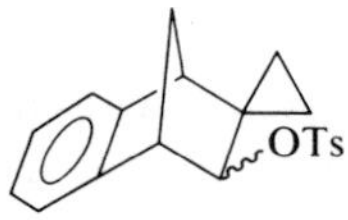

On the other hand, the tertiary derivative (**31**) gives high *exo* : *endo* rate ratio comparable to the values observed in the parent tertiary system (**30**)[33] and in the 2-anisyl derivative (**19**) (25). Consequently, if the basic

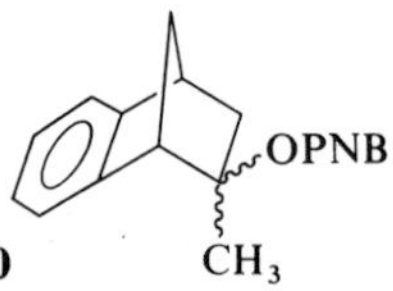

(25)

Exo : *endo*	6500	5200

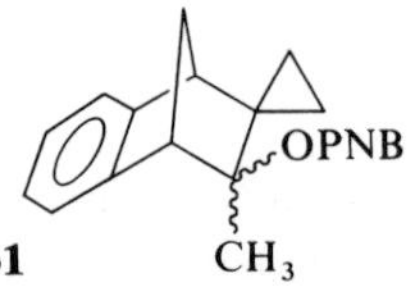

premise is correct, this must mean that in tertiary 2-methyl derivative, the very high *exo* : *endo* rate ratio cannot have its origin in π-participation.

There are some puzzling features about the secondary systems—they deserve a more detailed examination. First, it is to be observed that **26** is a secondary allylic system. Ordinarily such tosylates are too unstable to prepare. Yet these tosylates are not only readily prepared, but they undergo solvolysis at moderate rates. The stability is surprising.

[32]D. Lenoir, P. v. R. Schleyer, and I. Ipaktschi, *Ann.*, **750**, 28 (1971).
[33]H. C. Brown and G. L. Tritle, *J. Amer. Chem. Soc.*, **88**, 1320 (1966).

There is an even more disturbing feature about these derivatives. Despite their greatly reduced *exo* : *endo* reactivity ratio, both **26** and **27** are reported to yield *exo* acetate or alcohol in high yield—with no significant *endo* reported. This would appear not to fit the rate-product correlation represented by the Goering–Schewene diagram. Complete Goering–Schewene diagrams for these systems are greatly to be desired.

A possible interpretation of these unusually low *exo* : *endo* rate ratios both for the exchange of the ketal (**22**) and for these allylic and spirocyclopropyl-2-norbornyl derivatives (**25**) is discussed in the next section.

7.9. The Selectivity Principle and the 2-Norbornyl Problem

One of the problems that has hindered resolution of the 2-norbornyl question has been the failure of many workers to consider the "selectivity principle" and its possible effect on the interpretation of their results.[34]

The "selectivity principle" was introduced to account for the large amount of *meta* orientation (about 30%) observed in the Friedel–Crafts isopropylation of toluene. It has been considered anomalous that such a large amount of *meta* substitution should be observed in a benzene derivative containing an *ortho–para* directing substituent, such as methyl. It was then pointed out that this apparent anomaly had its origin in the neglect of the selectivity of the reaction.

Thus, bromination is a highly selective reaction and discriminates greatly between the *meta* (0.3% *meta*) and the *ortho + para* positions. Nitration is a reaction of intermediate selectivity, yielding 3.5% *meta* isomer. Isopropylation is a reaction of low selectivity, yielding 26% *meta* isomer.[35] If the toluene/benzene rate ratio is utilized as a measure of the reaction selectivity, one observes a value of 605 for bromination, 25 for nitration, and 1.8 for isopropylation. In other words, a reaction which exhibits a high degree of selectivity between toluene and benzene shows high selectivity between the *meta* and the *ortho + para* positions of toluene.[36]

It is a corollary of these considerations that in investigations of structural effects one should use reactions which are highly selective, that is, sensitive to structural features. If one uses reactions of low selectivity, it is evident that the effects of structural features can be severely dampened.

Let us select the cyclopentane and cyclohexane systems as suitable reference structures to estimate the relative selectivities of various reactions.

[34]H. C. Brown and E. N. Peters, *Proc. Nat. Acad. Sci. U.S.*, **71**, 132 (1974).
[35]L. M. Stock and H. C. Brown, *Adv. Phys. Org. Chem.*, **1**, 35 (1963).
[36]H. C. Brown and K. L. Nelson, *J. Amer. Chem. Soc.*, **75**, 6292 (1953).

On this basis, the acetolysis of tosylates at 25°C[37] is a reaction of reasonable selectivity:

(26)

RR (25°C):   32.4   1.00

Similarly, the solvolysis of *tert-p*-nitrobenzoates at 25°C[38] reveals reasonable selectivity:

(27)

RR (25°C):   38.5   1.00

These reactions also reveal high selectivity between the *exo* and *endo* isomers of norbornyl (28, 29).[8,39]

(28)

RR (25°C):   280   1.00

(29)

RR (25°C):   885   1.00

On the other hand, the decomposition of *tert*-butyl peresters reveals little selectivity between the cyclopentyl and cyclohexyl derivatives (30)[40]

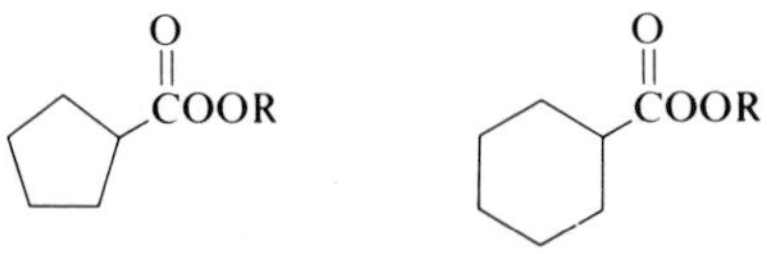

(30)

RR (25°C):   0.5   1.00

[37]S. Winstein, B. K. Morse, E. Grunwald, H. W. Jones, J. Corse, D. Trifan, and H. Marshall, *J. Amer. Chem. Soc.*, **74**, 1127 (1952).
[38]E. N. Peters and H. C. Brown, *J. Amer. Chem. Soc.*, **97**, 2892 (1975).
[39]S. Ikegami, D. L. Vander Jagt, and H. C. Brown, *J. Amer. Chem. Soc.*, **90**, 7124 (1968).
[40]P. Lorentz, C. Rüchardt, and E. Schacht, *Tetrahedron Lett.*, 2787 (1969).

and little selectivity between the *exo* and *endo* isomers of norbornyl (31).[41]

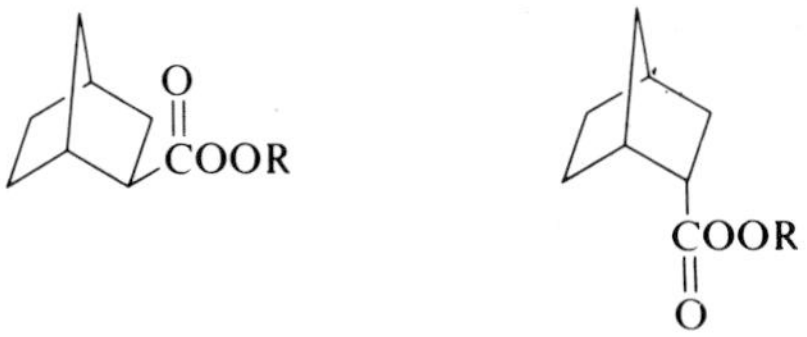

(31)

RR (60°C):   4.7   1.00

Clearly this last reaction is not a satisfactory one with which to study structural effects, either in the norbornyl system or in related systems.

With this factor in mind, let us examine those reactions which have been observed to give low *exo* : *endo* rate or product ratios and have been utilized to define the stereochemical properties of a "classical" 2-norbornyl cation. Are they reactions of reasonable selectivities, such as Reactions 26 and 27, or are they reactions of low selectivity, such as Reaction 30?

Unfortunately, the available data for the hydrolysis of ketals of cyclopentanone and cyclohexanone[42] were carried out under somewhat different conditions than the norcamphor study. However, the results (32) clearly indicate this to be a reaction of low selectivity.

(32)

RR (25°C):   3.4   1.00

The solvolysis of the 2-methylenecycloalkyl-3,5-dinitrobenzoates in aqueous dioxane at 100°C[43] likewise reveals a reaction of low selectivity

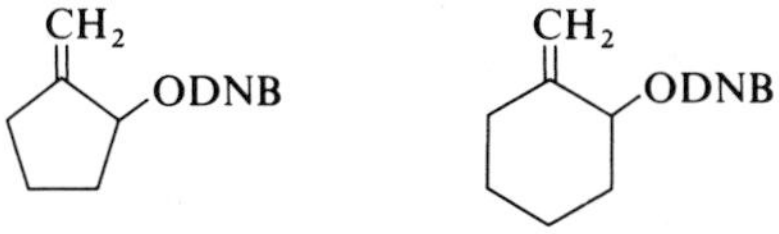

(33)

RR (100°C):   1.2   1.00

(33). This low selectivity carries over to the corresponding norbornyl derivatives (34).[31]

(34)

RR (125°C):   3.1   1.00
RR (OTs, 25°C):   3.9   1.00

[41]P. D. Bartlett, G. N. Fickes, F. C. Haupt, and R. Helgeson, *Accounts Chem. Res.*, **3**, 177 (1970).
[42]M. M. Krevoy, C. R. Morgan, and R. W. Taft, Jr., *J. Amer. Chem. Soc.*, **82**, 3064 (1960).

Finally, the solvolysis of the spirocyclopropane derivatives in aqueous dioxane[43] reveals that this reaction is also one of low selectivity (35). In this

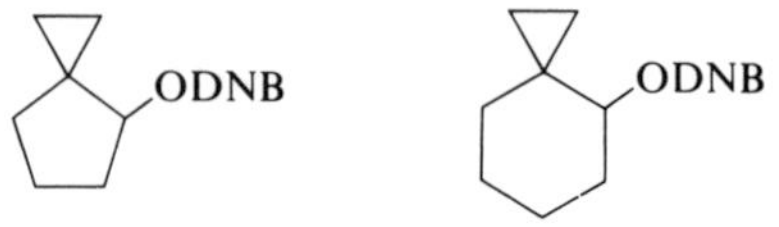

(35)

RR (100°C):   3.2   1.00

case also the low selectivity is reflected in the related norbornyl derivatives (36).[30]

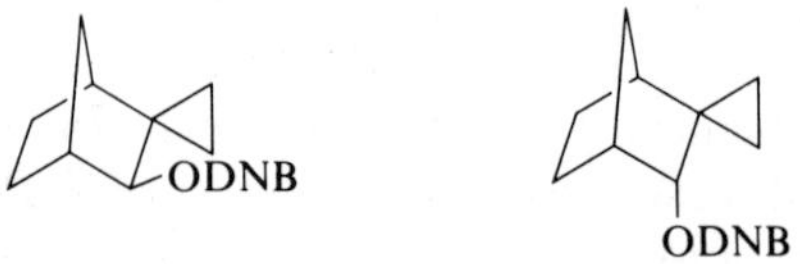

(36)

RR (100°C):   2.8   1.00

Consequently, it appears that those reactions, which have been used to arrive at the conclusion that the stereoselectivity of a "classical" 2-norbornyl system must be low, involve reactions of low selectivity. In the next chapter we shall show that many reactions reveal a high stereoselectivity in the norbornyl system, even reactions which do not involve carbonium ion intermediates.

7.10. Conclusion

We know of no evidence that the 2-*p*-anisyl-2-norbornyl and the 2-*p*-anisyl-2-camphenilyl cations are not classical cations without σ-bridges. Yet these two systems reveal high *exo* : *endo* rate and product ratios comparable to those in norbornyl itself. A comparison of the Goering–Schewene diagrams, Figures 6.3 and 6.4, with Figures 7.1 and 7.2 is impressive. If the difference in energy of the two transition states in Figures 7.1 and 7.2 cannot be the result of σ-participation, what can be responsible? In the next chapter we shall consider the possibility that it is the *endo* transition state which is unusual, destabilized by steric interactions.

[43]T. Tsuji, I. Moritani, and S. Nishida, *Bull. Chem. Soc. Jap.*, **40**, 2338 (1967).

Comments

Brown's discovery that *exo*: *endo* ratios remain large in *tertiary* 2-norbornyl systems presents a major challenge to the bridged-ion theory. His work also led to the now established position that tertiary 2-norbornyl cations are classical or nearly so. While partial bridging may be present in some of these species, this is unlikely to be significant energetically.

It certainly is possible that the factor (or factors) responsible for the high *exo*: *endo* ratios in tertiary 2-norbornyl systems might also contribute to the high secondary *exo*: *endo* ratios. However, nature need not be consistent in providing the same physical basis for the remarkably similar Goering–Schewene diagrams for secondary and tertiary 2-norbornyl systems. Indeed, quite different effects might be involved, and this is the conclusion I favor. Tertiary systems are not reliable models for secondary behavior; the steric effects associated with the former may largely be absent in the latter.

The low *exo*: *endo* ratios of the stabilized *secondary* systems **26** and **27** persuasively support the conclusion that the solvolysis of 2-*exo*-norbornyl tosylate is anchimerically asisisted.[44] Lenoir, Röll, and Ipaktschi[45] have now shown that the *exo*: *endo* ratio of the tertiary norbornyl esters (**32**) is also unchanged by the adjacent spirocyclopropane ring (**33**) (37).

		(80% acetone)	(60% acetone)
k 25°C sec^{-1}	*exo*	1.02×10^{-8}	2.71×10^{-3}
	endo	1.14×10^{-11}	3.26×10^{-6}
exo: *endo*		895	831

This result, parallel to that in the 2-benzonorbornenyl series (25), shows that such high *tertiary exo*: *endo* ratios are steric, rather than electronic in origin.

The behavior of secondary 2-norbornyl and 2-benzonorbornenyl systems are in sharp contrast. The high *exo*: *endo* ratios for the parents ($\sim 10^3$

---

[44]More detailed investigation of **26** has yielded an acceptable Goering–Schewene diagram, C. W. Jefford, private communication; for earlier work, see C. W. Jefford and W. Wojnarowski, *Helv. Chim. Acta*, **53**, 1194 (1970). The dinitrobenzoates of **26** yield *exo*: *endo* ratios comparable to the tosylates (34).

[45]D. Lenoir, W. Röll, and J. Ipaktschi, *Tetrahedron Lett.*, 3073 (1976).

for 2-norbornyl (28); 15,000 for 2-benzonorbornenyl (18)), are reduced to very low values (2.8 to 12) by the introduction of a spirocyclopropane ring (**27** and **29**) or a double bond (**26** and **28**) at the 3-position (24, 34, 36). I think that **26–29** provide an estimate of the *inherent* steric and hyperconjugative contributions to the *exo*:*endo* ratios in these *secondary* systems. On this basis, anchimeric assistance is responsible for the larger ratios found in 2-norbornyl (28) and 2-benzonorbornenyl (18).

"Selectivity" is unimportant here. Brown's arguments[34] in 7.9 have serious flaws. Because of the complications of solvent assistance during acetolysis, larger for cyclopentyl than for cyclohexyl,[46] it is better to use trifluoroacetolysis data (38).[47]

$$\text{(38)}$$

RR (TFA, 25°C):   13.8   1120

Brown's cyclopentyl/cyclohexyl comparison with *exo*:*endo*-2-norbornyl does not appear to be particularly apt, because the former is nearly 10^2 smaller in magnitude.

The following cyclopentyl/cyclohexyl ratios (all given, where possible, at 100°C, the temperature of measurement) do not provide convincing evidence for the operation of a selectivity effect [compare with (33), (35), and (38)].

11.4 (100°C, 60% acetone)[48]
9.5 (125°C, 80% acetone)[38]

17^{49}

13^{49} 2.0(140°C)[43]

[46]F. L. Schadt, T. W. Bentley, and P. v. R. Schleyer, *J. Amer. Chem. Soc.*, **98**, 7667 (1976).

[47]J. E. Nordlander, R. R. Gruetzmacher, W. J. Kelly, and S. P. Jindal, *J. Amer. Chem. Soc.*, **96**, 181 (1974); J. E. Nordlander, J. M. Blank, and S. P. Jindal, *Tetrahedron Lett.*, 3477 (1969). Also see J. B. Lambert and G. J. Putz, *J. Amer. Chem. Soc.*, **95**, 6313 (1973).

[48]R. C. Fort, Jr., R. E. Hornish, and G. A. Liang, *J. Amer. Chem. Soc.*, **92**, 7558 (1970).

[49]T. Tsuji, I. Moritani, S. Nishida, and G. Tadokoro, *Bull. Chem. Soc. Japan*, **40**, 2344 (1967).

In the secondary 2-norbornyl series, a 3-spirocyclopropyl group reduces the *exo*:*endo* ratio from 1120 (38) to 2.8 (36), or 400-fold. With similar substitution, the cyclopentyl/cyclohexyl ratios go from 13.8 (38) to 3.2 (35), only a 4-fold change!

Direct evidence argues against a selectivity effect on *exo*:*endo* ratios. If such an effect were present, it should be manifest with tertiary as well as with secondary norbornyl esters. This is not found. The 3-spirocyclopropane ring produces rate enhancements greater than 10^5, and yet the *exo*:*endo* ratio remains unaltered (37).[45] The same is true in the benzonorbornenyl series (25).[32]

For the sake of argument, let us use the selectivity relationship to estimate the extent of anchimeric assistance in 2-norbornyl. The benzyl derivatives show comparably large selectivity (39).

$$C_6H_5 \diagdown OPNB \Big/ C_6H_5 \diagdown OPNB \qquad \diagup\hspace{-0.5em}OPNB \Big/ \diagup\hspace{-0.5em}C_6H_5 \qquad (39)$$

RR (80% acetone, 25°C): 178^{50} 127^{51}

Trifluoroacetolysis is a reaction leading to classical or nearly classical carbocation transition states in the cyclopentyl and cyclohexyl series; selectivity decreases 13-fold (i.e., 178/13.8).

$$\diagdown OTs \Big/ \diagdown OTs$$

RR (TFA, 25°C):   13.8

This 13-fold decrease, combined with the data for the phenyl substituted norbornyl esters (39), yields an estimate of 10 (i.e., 127/13) for the *exo*:*endo* ratio of secondary norbornyl assuming *classical carbocation transition states are involved* (40).

$$\diagup\hspace{-0.5em}OTs \Big/ \diagup\hspace{-0.5em}OTs \qquad \begin{array}{l}\text{classical} \\ \text{models}\end{array} \qquad (40)$$

RR (TFA, 25°C):   10 (estimated)

[50]E. N. Peters and H. C. Brown, *J. Amer. Chem. Soc.*, **97**, 7454 (1975); H. C. Brown, M. Ravindranathan, and M. M. Rho, *ibid.*, **98**, 4216 (1976).

[51]H. C. Brown, K. Takeuchi, and M. Ravindranathan, *J. Amer. Chem. Soc.*, in print. This value has been revised slightly from that in the preliminary report, K. Takeuchi and H. C. Brown, *ibid.*, **90**, 2693 (1968).

The experimental value is 1120 (38). On this basis, a factor of 112 would be attributed to anchimeric assistance in 2-*exo*-norbornyl trifluoroacetolysis. The *exo:endo* ratio of 10 (40) is very close to that observed experimentally for classical secondary 2-norbornyl systems (34, 36).

Arguments of this type seem quite logical when presented, but should be regarded with healthy skepticism until enough cases are tested to probe their general validity.

8

Exo : Endo Rate Ratios
as a Steric Phenomenon

8.1. Introduction

The high *exo:endo* rate and product ratios in the solvolysis of 2-norbornyl derivatives has long been considered a unique characteristic that required a unique explanation—σ-bridging.[1,2] Now that it has been observed that reactions involving the highly stabilized 2-*p*-anisyl-2-norbornyl and 2-*p*-anisyl-2-camphenilyl cations exhibit equally high *exo:endo* rate and product ratios (Chapter 7), it is necessary to explore explanations other than σ-bridging as a basis for these characteristics.

8.2. Steric Characteristics of the Norbornyl System

At the time the high *exo:endo* rate and product ratios for the solvolysis of 2-norbornyl derivatives were first observed,[1] little was known about the remarkable steric characteristics of the norbornane structure. Consequently, it was considered that these high *exo:endo* rate and product ratios were an exceptional phenomenon requiring a special explanation—σ-bridging.

Since then, considerable additional data have become available bearing on the stereochemistry of reactions involving the norbornane structure. It has become evident that all reactions of norbornane exhibit a marked preference for reaction from the *exo* over reaction from the *endo* direction. The precise preference can vary considerably with the particular reaction

---

[1]S. Winstein and D. Trifan, *J. Amer. Chem. Soc.*, **74**, 1147, 1154 (1952).
[2]P. D. Bartlett, *Nonclassical Ions*, Benjamin, New York, 1965.

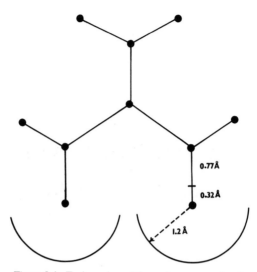

Figure 8.1. End-on view of the norbornane molecule.

utilized, but there appears to be no exception to the rule of preferential *exo* attack.

The origin of this preference for *exo* reaction appears to lie in the rigid U-shaped structure of the norbornane system (**1**). Carbon atoms 1–6

constitute a cyclohexane structure in the higher energy boat conformation. Moreover, the 7-methylene group not only locks this ring system into a rigid boat conformation, but the constraint so produced accentuates the steric crowding within the *endo* moiety of the boat structure (Figure 8.1). Consequently, the marked preference of all reactions for the *exo* face of the norbornane system is explicable on simple steric grounds.

To illustrate, free-radical chlorination of norbornane with sulfuryl chloride yields 95% *exo* chloride (1).[3]

$$\text{(norbornane)} + SO_2Cl_2 \xrightarrow{\text{peroxide}} \text{(norbornyl chloride)} + SO_2 + HCl \qquad (1)$$

95% *exo*

[3]P. D. Bartlett, G. N. Fickes, F. C. Haupt, and R. Helgeson, *Accounts Chem. Res.*, **3**, 177 (1970).

Similarly, hydroboration–oxidation of norbornene yields 99.5% *exo* product (2).[4]

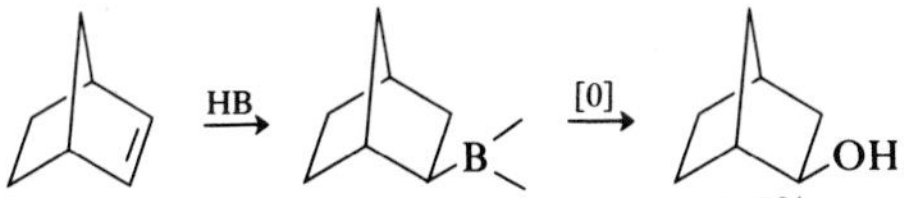

$$\text{(2)}$$

99.5% *exo*

Epoxidation of norbornene gives 99.5% of the *exo*-epoxide (3).[5]

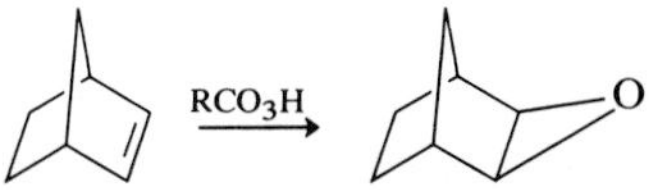

$$\text{(3)}$$

99.5% *exo*

Oxymercuration–demercuration of norbornene gives ≥99.8% *exo*-alcohol (4).[6]

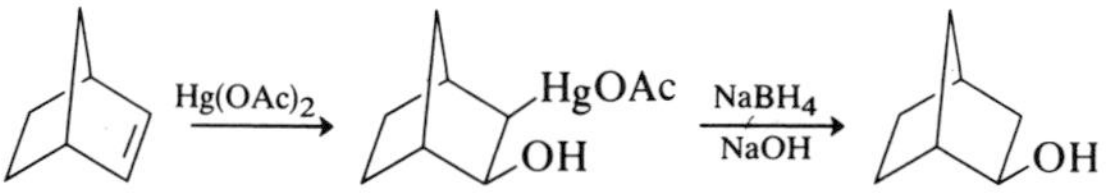

$$\text{(4)}$$

≥99.8% *exo*

Free-radical addition of thiophenol at 0°C gives 99.5% *exo*-thioether (5).[7]

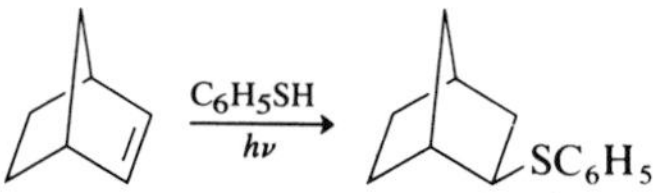

$$\text{(5)}$$

99.5% *exo*

Finally, the base-catalyzed deuterium exchange of norbornanone yields an *exo:endo* ratio of 715 (2).[8]

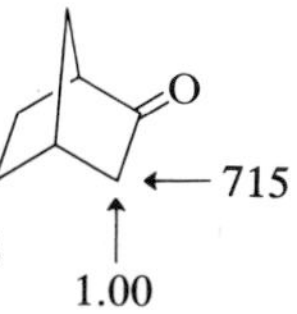

Therefore, the faster rate of solvolysis of the *exo*-norbornyl derivatives, both secondary and tertiary, and the *exo* stereochemistry of the solvolysis

[4]H. C. Brown and J. H. Kawakami, *J. Amer. Chem. Soc.*, **92**, 1990 (1970).
[5]H. C. Brown, J. H. Kawakami, and S. Ikegami, *J. Amer. Chem. Soc.*, **92**, 6914 (1970).
[6]H. C. Brown and J. H. Kawakami, *J. Amer. Chem. Soc.*, **95**, 8665 (1973).
[7]H. C. Brown, J. H. Kawakami, and K.-T. Liu, *J. Amer. Chem. Soc.*, **95**, 2209 (1973).
[8]T. T. Tidwell, *J. Amer. Chem. Soc.*, **92**, 1448 (1970).

products are not unique, but conform to the same reactivity pattern exhibited by the norbornyl system in varying degree in all of its reactions.

8.3. Steric Characteristics of the 7,7-Dimethylnorbornyl System

It was originally believed that 7,7-dimethyl substituents would dominate the stereochemistry of all reactions of the norbornyl system and force such reactions to take a preferential *endo* course in the absence of σ-bridging.[9] However, data now accumulating require modification of that assumption. Addition reactions that proceed through concerted cyclic processes are indeed dominated sterically by the *cis*-7-methyl substituent and directed preferentially to *endo* (6).[4]

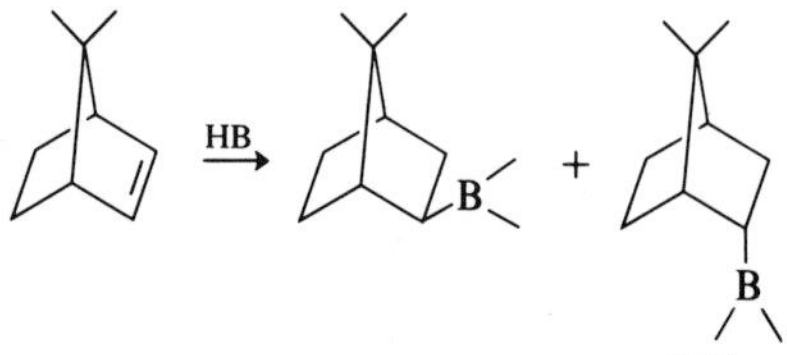

$$\text{(6)}$$

22%   78%

On the other hand, reactions that proceed through individual attack on C2 or C3 by species of not too large steric requirements are not dominated sterically by the *cis*-7-methyl substituent. Thus, the free-radical addition of thiophenol to 7,7-dimethylnorbornene proceeds to give the *exo*-thioether preferentially (7).[7]

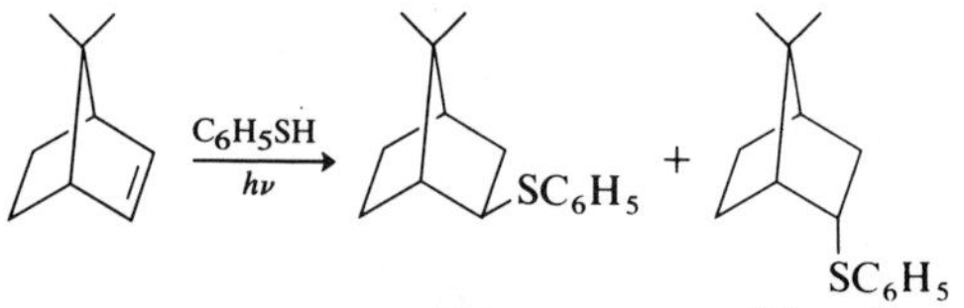

$$\text{(7)}$$

95% *exo*   5% *endo*

The base-catalyzed deuteration of camphor reveals an *exo*:*endo* ratio of 21 (3).[8]

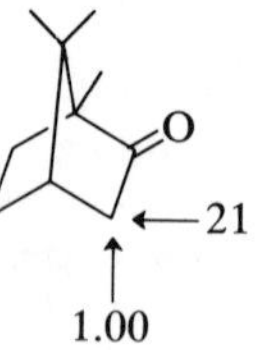

1.00

[9]J. A. Berson, *Molecular Rearrangements*, Vol. 1, P. de Mayo, Ed., Interscience, New York, 1963.

Numerous other reactions which involve attack by moieties of small steric requirements also go preferentially *exo* (8).[7]

$$\text{(8)}$$

>99.8% *exo*

The evidence favors the view that reactions involving reagents of very large steric requirements, such as the reduction of camphor by lithium aluminum hydride, or those proceeding through cyclic processes, such as the hydroboration (6) or epoxidation of 7,7-dimethylnorbornene, are forced by these substituents to go *endo*. However, many other reagents go preferentially *exo* (7, **3**, 8). Consequently, we can no longer consider it anomalous that 2-*p*-anisyl-2-norbornyl[10] and 2-*p*-anisyl-2-camphenilyl[11] derivatives undergo solvolysis to give *exo* products (Section 7.5).

8.4. Steric Hindrance to Ionization

The data reveal that the great majority of the reactions of the norbornyl system exhibit high *exo*: *endo* ratios. (In a few cases, such as the chlorination of norbornane with elementary chlorine and the decomposition of *tert*-butyl peresters, the ratios are small.[3] Presumably this is a consequence of the lower selectivity of these reactions, as discussed in Section 7.9.) These high ratios are presumably the result of decreased rates of reaction in the sterically hindered *endo* direction of the U-shaped norbornane structure (Figure 8.1). Consequently, it appeared appropriate to consider that the high *exo*: *endo* rate ratio in the solvolysis of the tertiary 2-*p*-anisyl-2-norbornyl and the parent secondary 2-norbornyl derivatives may actually be the result of a normal *exo* rate combined with a very slow *endo* rate.

The concept of steric assistance to ionization was introduced in 1946.[12] It was proposed that tertiary derivatives carrying bulky substituents would undergo solvolysis with relief of steric strain and would therefore exhibit enhanced rates (9).

$$\text{(9)}$$

tetrahedral   planar
(strained)   (less strained)

[10] H. C. Brown and K. Takeuchi, *J. Amer. Chem. Soc.*, **90**, 2691 (1968).
[11] K. Takeuchi and H. C. Brown, *J. Amer. Chem. Soc.*, **90**, 5268 (1968).
[12] H. C. Brown, *Science*, **103**, 385 (1946).

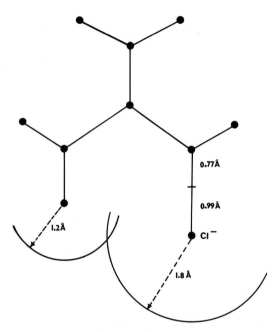

Figure 8.2. End-on view of the *endo*-norbornyl chloride molecule.

Considerable data has been accumulated to support this proposal (Chapter 2) and it appears to be generally accepted.[13]

The proposal of steric hindrance[14] to ionization has encountered much more resistance.[13] Yet it appears equally reasonable that the structure of the compound undergoing solvolysis may be so shaped that the departure of the leaving group may be hindered, as indicated in **4**.

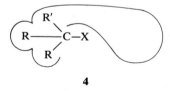

4

Indeed, *endo*-norbornyl derivatives would appear to possess this feature. Consider the *endo*-norbornyl chloride structure (Figure 8.2). In the ionization process the chlorine substituent would be expected to move along a curved path away from the carbon atom at the 2-position, maintaining the chlorine substituent perpendicular to the face of the developing carbonium

[13]W. J. le Noble, *Highlights of Organic Chemistry*, M. Dekker, New York, 1974.
[14]H. C. Brown, F. J. Chloupek, and M.-H. Rei, *J. Amer. Chem. Soc.*, **86**, 1248 (1964).

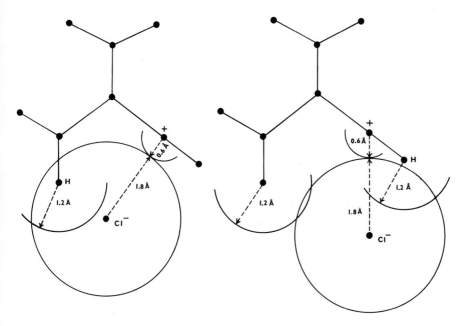

Figure 8.3. Molecular models for the hypothetical intimate ion pairs from the two postulated reaction paths.

ion so as to retain maximum overlap of the orbitals undergoing separation. In this way the system should pass through the transition state to the first intermediate, the idealized ion pair shown in Figure 8.3. Clearly, there would be a major steric overlap of the chlorine substituent with the *endo*-6-hydrogen. Moreover, the group undergoing ionization should be strongly solvated by the medium, yet the U-shaped structure obviously makes difficult such solvation of the developing anion.

An alternative model for ionization has been suggested.[15] In this model the departing group would move initially along the direction of the C—X bond leading to the first intermediate, the idealized ion pair shown in Figure 8.3. This path does not avoid the steric difficulty, although in this model it is transferred largely to the hydrogen atom or other group at the 2-position.

The large steric interactions of both models will presumably cause some other path, providing decreased steric interactions at the cost of poorer overlap, to be selected as a compromise. Such a compromise would still result in an increase in the energy of the transition state as compared to that for a derivative without this particular structural feature.

[15]P. v. R. Schleyer, M. M. Donaldson, and W. E. Watts, *J. Amer. Chem. Soc.*, **87**, 375 (1965).

8.5. Misconceptions

Let us consider some of the misconceptions and misunderstandings of this proposal which have appeared in the literature or been transmitted to me. The equilibration of *exo*- and *endo*-norbornyl derivatives, such as 2-norbornyl acetate,[16] yields equilibrium mixtures in which the *exo* isomer is usually favored by a small factor, in the neighborhood of 6 to 10. Thus, from the Goering–Schewene diagrams (Figures 6.3 and 6.4) the difference in the ground-state energies of the *exo*- and *endo*-isomers is 1.0 kcal mol^{-1} for the acetate and 1.3 kcal mol^{-1} for the arenesulfonate. If there is such a small difference in the ground-state energies, attributed to steric interactions, how is it possible to attribute differences of ~5 kcal mol^{-1} in the stabilities of the two transition states to steric interactions?

A careful examination of Figures 8.2 and 8.3 will clarify the situation. The equilibrium steric forces involve the *endo* substituent when it is in its usual position vertically downward from C2 (Figure 8.2). In the ionization process it moves inward, into the U-shaped cavity. Clearly, steric interactions will be increased. At the same time formation of the anion is generally accompanied by strong solvation. As the substituent moves into the cavity (Figure 8.3), such solvation must be seriously hindered. Clearly, the steric interactions must increase in going from the initial state (Figure 8.2) through the transition state to the ion pair (Figure 8.3).

A second problem appears to be the similar *exo*:*endo* ratios observed in 2-norbornenyl derivatives and 2-norbornyl derivatives. Does not the absence of the 6-hydrogen atom diminish the steric interactions?

The π-cloud of a double bond has been assigned a thickness equivalent to half the separation of the carbon layers in graphite.[17] In numerous reactions the effect of the π-electrons in controlling the direction of attack is as large or even larger than that of the ethano bridge.[18] Thus the reduction

$$\text{(10)}$$

by sodium borohydride (10) is more selective for dehydronorcamphor than for norcamphor.[18] The situation is rendered clear by Figure 8.4.

---

[16]H. L. Goering and C. B. Schewene, *J. Amer. Chem. Soc.*, **87**, 3516 (1965).

[17]L. Pauling, *The Nature of the Chemical Bond*, 3rd ed., Cornell University Press, Ithaca, N.Y., 1969.

[18]H. C. Brown and J. Muzzio, *J. Amer. Chem. Soc.*, **88**, 2811 (1966).

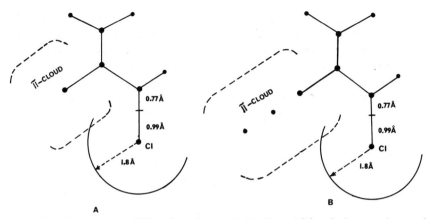

Figure 8.4. End-on views of (A) *endo*-norbornenyl chloride and (B) *endo*-benzonorbornenyl chloride.

"If Brown's picture is to be used as a guide, one should expect enormous rate retardations for such compounds as

But in fact the rate retardation for . . . [this compound is] only a factor of 3" [Reference to] P. von R. Schleyer, as quoted by K. B. Wiberg and G. R. Wenzinger, *J. Org. Chem.*, **30**, 2278 (1965) (Table V, footnote *d*).[13]

The fact is that the rate for this compound relative to *endo*-norbornyl is 0.1. But one should not compare raw rates directly for systems involving major differences in strain energies. If one applies the Foote–Schleyer correlation here, the predicted rate is 1000 times greater than it is for *endo*-norbornyl. Consequently, the observed rate is slower than the predicted value by a factor of 10,000!

Similarly, le Noble[13] points out that the brosylate **5** solvolyzes 300 times faster than the brosylate (**6**).[19] Should not steric retardation be anticipated

OBs

5   **6**

[19]P. Carter and S. Winstein, *J. Amer. Chem. Soc.*, **94**, 2171 (1972).

for **5**? Unfortunately, it is no longer possible to give a glib answer. In simple aliphatics steric strain always decreases as the leaving group departs (9). In these bicyclics one must estimate the strain in the ground state and the strain in the transition state to arrive at a decision as to whether the reaction will be accelerated or retarded.

The fusion of a benzo group in **7** is such as to avoid the greatly increased ground-state energy of **5**, while actually enhancing the U-shaped character

RR (25°C):   0.00072   0.92
endo-norbornyl = 1.00

of the *endo* cavity.[20] The system reveals an *exo*: *endo* rate ratio of 1270, not because of any high rate for the *exo* isomer. (It actually undergoes solvolysis at a rate lower than *endo*-norbornyl tosylate.) The high *exo*: *endo* rate ratios arise primarily from the very low rate for the *endo* isomer.

The Foote–Schleyer correlation was briefly discussed earlier (Section 6.9). Perhaps a more detailed discussion will clarify the situation.

8.6. Steric Hindrance to Ionization and the Foote–Schleyer Correlation

If it were possible to calculate the rate for *endo*-norbornyl tosylate, without allowing for steric hindrance to ionization, agreement of the calculated with the observed value would confirm the absence of such a steric strain term. Then the faster rate for the *exo* derivative would support the proposal that this transition state is stabilized by major σ-participation. Alternatively, if the calculated rate for the *exo* agreed with the observed value, σ-participation could not be a factor and we would be forced to consider a steric explanation for the slower *endo* rate.

The Foote–Schleyer correlation was proposed as a means of making such a calculation (11)[21,22]:

$$\log k_{rel} = \frac{1715 - \nu_{CO}}{8} + 1.32 \sum_i (1 + \cos 3\phi_i) + \text{inductive term} + \frac{GS_{strain} - TS_{strain}}{1.36} \quad (11)$$

[20]R. Baker and J. Hudec, *Chem. Comm.*, 929 (1967).
[21]C. S. Foote, *J. Amer. Chem. Soc.*, **86**, 1853 (1964).
[22]P. v. R. Schleyer, *J. Amer. Chem. Soc.*, **86**, 1854, 1856 (1964).

This correlation utilizes the infrared carbonyl frequency of the ketone related to the secondary tosylate[23] under consideration as a method of estimating the contribution of angle strain to the solvolysis rate. The second term correlates for torsional effects. The third term corrects for the inductive effect of substituents, but is not required in the present discussion. The last term is important for the present discussion. It represents the effect of differences in the ground (GS) and transition states (TS) upon the relative rate. In compounds where such nonbonded strains were not considered to be serious, good agreement was realized between calculated and observed rates.[22] In this discussion we will consider several cases where nonbonded interactions involving the substituent could be quite serious, if steric hindrance to ionization is a real phenomenon.

In principle, the nonbonded strain term of the correlation should be capable of handling both steric assistance to ionization ($GS_{strain} > TS_{strain}$ for the leaving group) and steric hindrance to ionization ($GS_{strain} < TS_{strain}$). However, in practice it was recognized that the latter situation would be difficult to calculate, since neither the path of departure of a leaving group nor the proper nonbonded potential functions are known. Consequently, for all cases the simplest assumption was adopted, that $TS \approx 0$ for the leaving group.

This simplifying assumption was justified on the basis that the steric environment of the leaving group in the six model compounds used to define the correlation line, as well as in *endo*-norbornyl, was similar to the steric environment of a leaving group in an axial position of the cyclohexane molecule. In such flexible cyclohexane structures there is actually no evidence of any appreciable steric hindrance to ionization.[24] Accordingly, the assumption of $TS \approx 0$ was justified on the ground that "steric deceleration, in fact, is but rarely encountered, evidently because leaving groups are generally able to find a propitious avenue for departure."[22]

A major flaw in this argument may arise from the extrapolation from flexible alicyclic ring systems to rigid bicyclic structures. Steric effects are generally relatively small in flexible systems. The system can bend or rotate in such a manner as to minimize steric interactions. In rigid bicyclic systems, this avenue of escape is largely absent. Steric effects can be much larger[18] (Section 2).

---

[23]The correlation was originally restricted to the acetolysis of secondary tosylates upon the accepted position at that time that the acetolysis of such systems was close to limiting. Consequently, it does not contain a term for the contributions of solvent to the rate. This position is now undergoing reconsideration; see J. L. Fry, C. J. Lancelot, L. K. M. Lam, J. M. Harris, R. G. Bingham, D. J. Raber, R. E. Hall, and P. v. R. Schleyer, *J. Amer. Chem. Soc.*, **92**, 2538 (1970), footnote 21. Fortunately, the acetolysis of bicyclic tosylates of the norbornyl type, under examination here, is limiting or close to it.

[24]E. L. Eliel, N. L. Allinger, S. J. Angyal, and G. A. Morrison, *Conformation Analysis*, Wiley, New York, 1965, pp. 84–85.

Accordingly, it was decided to test the generality of the assumption that GS steric strain involving the leaving group can be assumed to vanish in the transition state.[25] The 6,6-dimethylnorbornane and the *endo*-5,6-trimethylenenorbornane systems exaggerate the U-shaped structural feature of the *endo* side of the norbornane molecule. Consequently, the molecules, 6,6-dimethyl-*endo*-norbornyl tosylate (**9**), *endo*-5,6-trimethylene-*endo*-2-norbornyl tosylate (**10**), and *endo*-5,6-trimethylene-*endo*-8-norbornyl tosylate (**11**) were selected for study (12).

$$\text{(12)}$$

The corresponding 9-derivative (**12**) was included to provide a test of the Foote–Schleyer correlation for a less rigid system. In this case the flexibility at the 9-position, greater than at the 8-, should serve to minimize steric hindrance to the departure of the leaving group.

Obviously, if steric hindrance to ionization is a factor, it should be much more important in the *endo*-2 (**10**) and *endo*-8 (**11**) derivatives than in *endo*-norbornyl tosylate. If it were possible to estimate the rates of **9**, **10**, and **11** and similar molecules with the assumption of TS ≈ 0 (for strain involving the leaving group), the steric hindrance to ionization must be negligible in *endo*-norbornyl tosylate and the Foote–Schleyer correlation could be used with considerable confidence to resolve the nonclassical ion problem, as already proposed.[22] On the other hand, if the observed rates are much slower than those calculated in this manner, the assumption that TS ≈ 0 cannot be universally correct. Such a result would not necessarily show that TS > 0 in the *endo*-norbornyl system, but it would lower the level of assurance that TS can be considered to be negligible in the solvolysis of *endo*-norbornyl tosylates and related rigid bicyclic derivatives.

The carbonyl frequency of the ketone corresponding to **10** is 1743 cm^{-1}, as compared to 1751 cm^{-1} for 2-norbornanone itself. According to the correlation (11), a shift of this magnitude should correspond to an increase in the rate of acetolysis of **10** over *endo*-norbornyl by a factor of 10. Differences in torsional and inductive effects between **10** and norbornyl are negligible. In the ground state, the strain in *endo*-norbornyl tosylate involving the leaving group is estimated as 1.3 kcal mol^{-1}. That in *endo*-2 (**10**) is estimated as 4.0 kcal mol^{-1}. On the original assumption of Schleyer that steric strain is largely relieved in the transition state, this will result in

[25]H. C. Brown, I. Rothberg, P. v. R. Schleyer, M. M. Donaldson, and J. J. Harper, *Proc. Nat. Acad. Sci. U.S.*, **56**, 1653 (1966).

another factor of 100 favoring *endo*-2 (**10**) over *endo*-norbornyl tosylate. Consequently, the Foote–Schleyer correlation, with the usual assumption that $TS_{strain} \approx 0$ for the leaving group, predicts that *endo*-2 (**10**) will solvolyze at a rate 1000 times that of *endo*-norbornyl tosylate.

The observed experimental relative rate of acetolysis is 0.1. Thus there is a discrepancy of 10,000 between the predicted and observed relative rates!

In 6,6-dimethyl-*endo*-norbornyl (**9**) the discrepancy is 8000!

For **11** and **12** the cyclopentyl system is selected as standard. The *endo*-8 (**11**) derivative gives a discrepancy of 100,000! On the other hand, in the case of *endo*-9 (**12**), where steric hindrance to ionization is not expected to be a significant factor, the calculated and observed rates of acetolysis agree within a factor of 2.

The data are summarized in Table 8.1.

These results point to an important conclusion. Despite the large ground-state strain present in U-shaped structures examined relative to the model substances used for comparison, solvolysis involves not a decrease, but an actual increase in strain in proceeding to the transition state, producing not an increase but an actual decrease in rate. It is clear that it will be necessary in the future to give careful consideration to the precise model for the departure of the leaving group.

TABLE 8.1. *Calculated and Observed Rates of Acetolysis at 25°C*

Tosylate	ν_{CO} for ketone, cm^{-1}	GS,[a] kcal mol^{-1}	$10^8 k_1$, sec^{-1} at 25°C	Calcd.	Obsd.	Discrep.	TS,[b] kcal mol^{-1}
endo-Norbornyl	1751	1.3	8.28		1.00		
endo-2-[d] (**10**)	1743	4.0	0.860	1000	0.10	10,000	6.8
6,6-Dimethyl-*endo*- (**9**)	1746	4.0	0.447	420	0.054	8,000	6.6
Cyclopentyl	1740	0.0	158		1.00		
endo-9-[d] (**12**)	1739	0.0	71.9	1.0	0.46	2	
exo-9-[d]	1739	0.0	44.8	1.0	0.28	3	
endo-8-[d] (**11**)	1732	3.0	1.56	1000 $(200)^c$	0.0099 $(0.18)^c$	100,000 $(1,000)^c$	7.2 $(4.1)^c$
exo-8-[d]	1732	0.0	8.88	10	0.056 $(1.0)^c$	180	

[a] Ground-state strain associated with the leaving group. This is the $(GS - TS_{strain})$ term of the Schleyer correlation, with the assumption that for the leaving group $TS_{strain} \approx 0$.
[b] Value of leaving group TS required in the $(GS - TS)$ term to achieve agreement between the calculated and observed relative rates.
[c] Values in parentheses are based on *exo*-8 as a model for *endo*-8.
[d] *endo*-5,6-Trimethylene-*exo*- or -*endo*-(2, 8, or 9)-norbornyl tosylate.

8.7. Steric Effects in U-Shaped Systems

The results described in the previous section encouraged the belief that steric hindrance to ionization could be a major factor in the observed *exo*:*endo* rate ratios in such systems. However, this represents a major revolution in chemical thought. For many years workers had been almost automatically attributing high *exo*:*endo* rate ratios to σ-participation in the *exo* isomer.

How might this idea be tested further?

One promising approach appeared to be to explore more widely the apparent relationship between the high *exo*:*endo* rate ratio in the norbornyl system and the high *exo*:*endo* product ratio in the norbornyl system for many nonsolvolytic reactions, such as hydroboration–oxidation, epoxidation, oxymercuration–demercuration, etc. Accordingly, we decided to examine the stereochemistry of selected reactions of a series of three bicyclic systems of increasing U-shaped character, and *exo*:*endo* rate ratios in solvolysis. The systems selected for study, in order of increasing U-shaped character, were *cis*-bicyclo[3.3.0]octane < norbornane < *endo*-5,6-trimethylenenorbornane (13).[26] We utilized the olefins (13, 14, and 15) and the ketones (16, 17, and 18).

$$(13)$$

The results are summarized in Table 8.2.

Although individual reactions evidently differ considerably in the stereoselectivities they exhibit, the results reveal a consistent pattern. In all cases, the *cis*-bicyclo[3.3.0]octane system, 13 and 16, exhibits the least preference for *exo* attack, presumably because of its higher flexibility and relatively less inaccessible *endo* face, whereas the *endo*-5,6-trimethylenenorbornane system exhibits the highest stereoselectivity of the three systems examined. Indeed, an examination of a molecular model

[26]H. C. Brown, W. J. Hammar, J. H. Kawakami, I. Rothberg, and D. L. Vander Jagt, *J. Amer. Chem. Soc.*, **89**, 6381 (1967).

reveals that in this rigid U-shaped structure the *endo* face is highly hindered to the approach of reagents. In all cases, the stereoselectivity indicated by the norbornane system is intermediate.

The corresponding tertiary methyl and tertiary phenyl *p*-nitrobenzoates (**19, 20, 21**) were then synthesized and the rates of solvolysis in 80% aqueous acetone determined.

The tertiary methyl ester from the bicyclo[3.3.0]octane system (**19**, R = Me) reveals an *exo* : *endo* ratio of 17. The norbornyl system (**20**, R = Me) reveals an *exo* : *endo* rate ratio of 885. Finally, the *endo*-5,6-trimethylenenorbornane derivative (**21**, R = Me) reveals an *exo* : *endo* rate ratio of 4300.

TABLE 8.2. *Comparison of the Relative Stereoselectivities Exhibited by Three Representative U-Shaped Systems*

| | *exo* : *endo* ratios for the U-shaped systems | | |
| | | | |
Reaction	13, 16, 19	14, 17, 20	15, 18, 21
Hydroboration–oxidation of olefin	24	200	>1,000
Epoxidation of olefin	6.7	200	>1,000
Oxymercuration–demercuration of olefin	8	>500	
Lithium aluminum hydride reduction of ketone	3	8.1	>1,000
Addition of CH$_3$MgX to ketone	50	200	>1,000
Oxymercuration of methylene derivatives	8.1	200	>1,000
Solvolysis of tertiary methyl *p*-nitrobenzoates	17	885	4,300
Solvolysis of tertiary phenyl *p*-nitrobenzoates	10	143	11,000

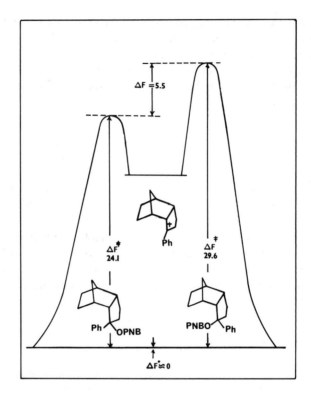

Figure 8.5. Free-energy diagram for the solvolysis of the 8-phenyl-*endo*-5,6-trimethylene-8-norbornyl *p*-nitrobenzoates in 80% acetone at 25°C.

The corresponding tertiary phenyl derivatives exhibit the same pattern of reactivity.[27] It is of special interest to note the similarity in the Goering–Schewene diagram for the solvolysis of 8-phenyl-*endo*-5,6-trimethylene-8-norbornyl *p*-nitrobenzoate (Figure 8.5) with the corresponding diagrams for the 2-*p*-anisyl-2-norbornyl (Figure 7.1) and 2-*p*-anisyl-2-camphenilyl (Figure 7.2) derivatives.

The *exo*:*endo* ratio remains high, 280, for the secondary 2-norbornyl tosylates. However, it drops down to 6 for the secondary *endo*-8 tosylates (**11**) (Table 8.1). This discrepancy will be discussed later when we undertake to extrapolate these results to the secondary system (Chapter 11).

The above results establish the existence of an excellent correlation within these three systems as to the stereoselectivities they show toward the various representative reagents and the stereoselectivities they reveal in the solvolysis of the related tertiary esters. This common pattern of reactivities

[27]H. C. Brown and D. L. Vander Jagt, *J. Amer. Chem. Soc.*, **91**, 6848, 6850 (1969).

for the carbonium ion and noncarbonium ion reactions supports the position we have been led to, that is, the *exo:endo* rate and product ratios in norbornyl must be largely, if not entirely, steric in origin.

8.8. Exo:Endo Rate and Product Ratios as a Steric Phenomenon

It is appropriate at this time to undertake a more detailed consideration of the effect of introducing 6,6- and 7,7-dimethyl substituents into 2-methylnorbornyl on the *exo:endo* rate ratio and the products produced in the solvolysis. The data presented lead to the conclusion that in the tertiary norbornyl derivatives we have examined the *exo:endo* rate ratio must be predominantly the result of steric forces. If this interpretation is valid, we should be able to introduce substituents into the *exo* face or the *endo* face of the norbornyl system and vary both the steric environment and the *exo:endo* rate ratio in a consistent, predictable manner.[28,29]

For example, the *exo:endo* rate ratio in the solvolysis of the 2-methyl-2-norbornyl *p*-nitrobenzoates (**22, 23**) in 80% acetone is 885. The steric requirements of the methyl and *p*-nitrobenzoate groups are very similar.[30]

	22 CH$_3$	**23** OPNB
RR (25°C):	885	1.00

Consequently, the ground-state energies for the *exo*-**22** and *endo*-**23** isomers are almost the same (Figure 8.6). The high *exo:endo* rate ratio is then ascribed to the fact that the *exo* face is more open sterically than the *endo* face. The greater steric availability of the *exo* face facilitates solvation of the incipient anion and its more ready ionization to the ion pair as compared to the more crowded steric environment of the *endo* face.

On this basis, an increase in the steric requirements of the *exo* face, while maintaining constant those of the *endo* face, should result in a major decrease in the *exo:endo* rate ratio. Contrariwise, maintaining the steric requirements of the *exo* face constant while increasing those of the *endo* face should bring about a significant increase in the *exo:endo* rate ratio.

[28]H. C. Brown and S. Ikegami, *J. Amer. Chem. Soc.*, **90**, 7122 (1968).
[29]S. Ikegami, D. L. Vander Jagt, and H. C. Brown, *J. Amer. Chem. Soc.*, **90**, 7124 (1968).
[30]M.-H. Rei and H. C. Brown, *J. Amer. Chem. Soc.*, **88**, 5335 (1966).

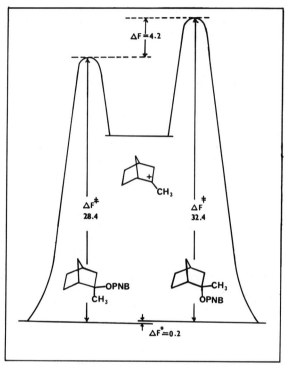

Figure 8.6. Free-energy diagram for the solvolysis of the 2-methyl-2-norbornyl *p*-nitro-benzoates in 80% acetone at 25°C.

Indeed, the introduction of *gem*-dimethyls into the 7-position decreases the *exo*:*endo* rate ratio from the 885 value for the parent compounds **22**:**23** to a value of 6.1 in **24**:**25**.

RR (25°C):   6.1   1.00

On the other hand, the introduction of *gem*-dimethyl into the 6-position increases the *exo*:*endo* rate ratio in **26**:**27** to 3,630,000.

RR (25°C):   3,630,000   1.00

Thus the *exo:endo* rate ratio changes by a factor of 600,000 merely through a shift of the methyl substituents from the 7- to the 6-position—a truly remarkable effect. As discussed earlier, a similar increase in the *exo:endo* rate ratio occurs in 2-*p*-anisyl-2-norbornyl with the introduction of methyl substituents in the 3-position (Section 7.4). Presumably, the *endo*-3-methyl group enhances the U-shaped cavity which traps the departing anion.

The free-energy diagram for the two trimethyl-2-norbornyl systems are given in Figures 8.7 and 8.8.

A comparison of the three diagrams, Figures 8.6–8.8, is instructive. The difference in the energies of the two transition states changes from 1.6 to 4.2 to 7.1 kcal mol^{-1}. Thus the ions produced in the solvolysis of each epimeric pair will distribute themselves to product in a ratio determined by this quantity. The product from the solvolysis of these compounds correspond to those predicted from the free-energy diagrams. Thus the alcohol product from **26** is exclusively ($\geq$99.9%) *exo*, whereas that from **24** contains ~10% *endo*.

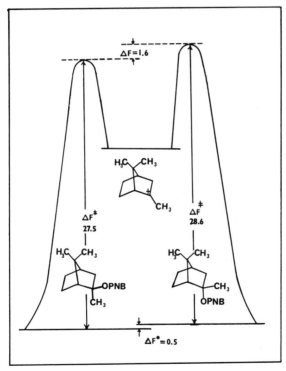

Figure 8.7. Free-energy diagram for the solvolysis of the 2,7,7-trimethyl-2-norbornyl *p*-nitrobenzoates in 80% acetone at 25°C.

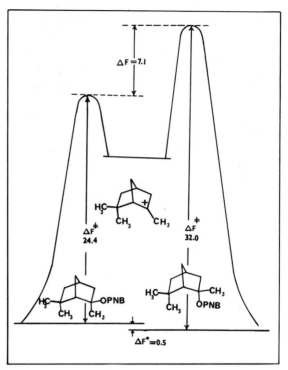

Figure 8.8. Free-energy diagram for the solvolysis of the 2,6,6-trimethyl-2-norbornyl
p-nitrobenzoates in 80% acetone at 25°C.

An examination of the actual rate constants (Table 2.2) is instructive. In
this discussion the rate constants will be compared to 1-methylcyclopentyl
p-nitrobenzoate (**28**) as standard.

	22 CH$_3$	28 CH$_3$	23 OPNB
RR (25°C):	4.74	1.00	0.00536

It should be noted that the rate constant for 2-methyl-*exo*-norbornyl
(**22**) is modestly greater than that for 1-methylcyclopentyl (**28**). However,
the *endo* isomer (**23**) is much slower. This is consistent with the proposed
interpretation in terms of steric hindrance to ionization.

The low *exo:endo* rate ratio in the 2,7,7-trimethylnorbornyl system
arises from a small increase in the rate of the *exo* isomer (**24**) and a large

increase by a factor of 580 in the rate of the *endo* isomer (**25**).

	24	**28**	**25**
RR (25°C):	19.0	1.00	3.1

Obviously, this marked increase is readily accounted for in terms of relief of steric strain accompanying rotation of the 2-methyl substituent away from the *gem*-dimethyl group during the ionization process (**29, 30**).

The high *exo:endo* rate ratio in the 6,6-dimethyl derivative comes about through a major increase in the rate of the *exo* isomer and a very slow rate for the *endo* isomer (**26, 27**).

	26	**28**	**27**
RR (25°C):	3440	1.00	0.000948

The very fast rate of 2,6,6-trimethyl-*exo*-norbornyl *p*-nitrobenzoate (726 greater than that of the parent compound, **22**) is again attributed to relief of steric strain (**31**).

As was pointed out earlier, the steric requirements of a methyl and *p*-nitrobenzoate group are quite similar.[30] Consequently, **32** should be just as strained as **31**. Yet **32** solvolyzes not faster, but slower than the much less strained parent compound **29**. This is readily accounted for on the basis that the natural reaction path for ionization of the *p*-nitrobenzoate group does not allow for relief of steric strain.

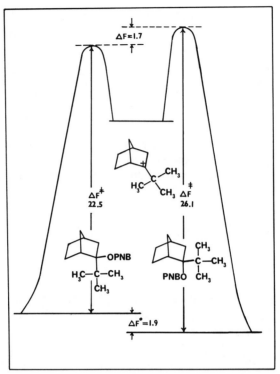

Figure 8.9. Free-energy diagram for the solvolysis of the 2-*tert*-butyl-2-norbornyl *p*-nitro-
benzoates in 80% acetone.

The importance of steric factors in these rigid bicyclic systems is
further illustrated by the behavior of the 2-*tert*-butyl-2-norbornyl *p*-
nitrobenzoates.[31] The *exo* : *endo* rate ratio (**33** : **34**) is 470, not greatly
different from the 885 ratio for the methyl derivative (**22** : **23**).

RR (25°C):   470   1.00

The introduction of the 2-*tert*-butyl group has a large effect on the
relative ground-state energies of the *exo* and *endo* isomers. This results in a
free-energy diagram (Figure 8.9) which reveals a relatively small difference,

[31]E. N. Peters and H. C. Brown, *J. Amer. Chem. Soc.*, **96**, 265 (1974).

1.7 kcal mol^{-1}, in the energies of the two transition states. This difference predicts an *exo:endo* product ratio of 17.6; a value of 19 was observed.

8.9. Conclusion

As a result of these many studies, it has become evident that the behavior of such tertiary 2-norbornyl derivatives can be clearly understood in terms of the effect of steric interactions, with remarkable consistency, without the need to postulate σ-bridging. The question remaining is whether the results can be extended to secondary norbornyl derivatives or is σ-bridging still required to understand the behavior of these secondary derivatives.

Comments

Brown has convincingly demonstrated that high norbornyl *exo*: *endo* rate and product ratios per se do not require a bridged-ion explanation. Many reactions not involving carbocations exhibit high ratios due to a combination of steric, torsional,[32] and nonequivalent orbital extension[33] (or other electronic) effects. On the other hand, many reactions give low *exo*: *endo* ratios, including the majority of processes involving 2-norbornyl free radicals.[3]

It would seem that each chemical process has its own characteristics. Thus, it is dangerous to attempt to model one reaction by another, until a direct relationship between them has been unequivocally established. The same reservation holds true for Brown's comparison between tertiary and secondary carbocation systems. Solvolysis, even of related secondary and tertiary esters, may be quite different mechanistically. There is no dispute that secondary systems are, in general, more prone to both neighboring group and solvent participation while tertiary systems, in general, are sterically more encumbered and less susceptible to such participation.

Occasionally, the *exo*: *endo* solvolysis ratios of tertiary and secondary esters are comparably large. Such cases are emphasized by Brown, but they seem to me to be only fortuities in view of abundant contrary examples. Consider the following systems, discussed in this Chapter.

	11 H OTs	21 CH$_3$ OPNB	Ph OPNB
exo: *endo*:	5.7	4300	11,000

The data for the 8-substituted *endo*-5,6-trimethylenenorbornyl systems (**11** and **21**) demonstrate that tertiary *exo*: *endo* ratios can be high, but if the solvolysis of the corresponding secondary esters are achimerically unassisted, a low ratio can be found. Brown terms this case a "discrepancy," but he does not provide an explanation.

The following secondary and tertiary *exo*: *endo* 2-norbornyl ratios respond oppositely when a *gem*-dimethyl group is attached to the 6-position (**35** vs. **26**, **27**) or to the 7-position (**36**[34] vs. **24**, **25**).

[32]P. v. R. Schleyer, *J. Amer. Chem. Soc.*, **89**, 699, 701 (1967).
[33]S. Inagaki and K. Fukui, *Chem. Lett.*, 509 (1974); S. Inagaki, H. Fujimoto, and K. Fukui, *J. Amer. Chem. Soc.*, **98**, 4054 (1976).
[34]For data, see Table 11.4.

H₃C CH₃ structures (chemical diagrams)

exo:endo (HOAc, 25°C): 280   206   4100

35 CH₃ OTs   **36** OTs

exo:endo (25°C):   885   3,630,000   6.1

22,23 OPNB   **26,27** CH₃ OPNB   **24,25** OPNB

That tertiary esters are much more sensitive to steric effects is also illustrated by comparing (14)[34] with (15):

$$\text{(14)}$$

RR (25°C):   1.0   0.75

$$\text{(15)}$$

RR (25°C):   1.0   580

Such disparate results are commonly encountered. Unfortunately, I can only conclude that tertiary data cannot be relied upon to model and to interpret the solvolysis rates of anchimerically unassisted secondary systems.

Brown terms the norbornyl system a "U-shaped structure" and emphasizes the steric consequences to be expected. He appreciates but does not emphasize that simple norbornyl derivatives are not very crowded in the ground state. For example, Figures 6.3 and 6.4 show that 2-*endo*-esters are only about 1 kcal/mole less stable than their 2-*exo* counterparts. The *axial-equatorial* energy differences for corresponding cyclohexyl derivatives are comparable in magnitude, and this is general behavior.[35] The introduction of methyl groups onto the norbornane framework does not introduce appreciable strain (16).[36] It is not appropriate to regard such simply substituted norbornanes as being crowded systems. If steric effects are manifest during chemical reactions, crowded *transition states* must be involved, or truly bulky substituents must be present.

(16)

Calculated strain energy, kcal/mole:	16.9	17.1	18.0

Recently, Menger and Thanos[37] examined the behavior of the *exo*- and *endo*-2-norbornyldimethylamines **37** and **38**. They concluded that these "compounds do not display substantial differences as would be expected if the *endo*-dimethylamino group were subjected to unusual steric or solvation effects within the endo cavity."

37   **38**

I agree with Brown that the use of the Foote–Schleyer equation to interpret the behavior of crowded systems is dangerous, since there is no way at present of predicting whether ground-state leaving group strain will be relieved or will be increased on ionization. However, no one considers 2-*exo*-norbornyl tosylate to be crowded, and this reservation is not applicable here.

Let me restate.the problem in simple terms: The free energy diagrams show that ground-state energy differences between *exo* and *endo* derivatives generally are small. The *ions* obtained from both are the same. Thus, neither ground states nor intermediates can be responsible for the *exo*:*endo* ratios observed. Differences in *exo* and *endo transition states* must be responsible.

[35]See, e.g., R. J. Ouellette, J. D. Rawn, and S. N. Jreissaty, *J. Amer. Chem. Soc.*, **93**, 7117 (1971).
[36]E. M. Engler, J. D. Andose, and P. v. R. Schleyer, *J. Amer. Chem. Soc.*, **95**, 8005 (1973).
[37]F. M. Menger and T. E. Thanos, *J. Amer. Chem. Soc.*, **98**, 3267 (1976).

Steric hindrance to ionization is a plausible explanation for the behavior of tertiary 2-norbornyl systems. The locus of departute of an *endo* leaving group can be restricted by a 2-*exo*-methyl or phenyl substituent. I doubt if a much smaller hydrogen can function similarly. Another possible difference should be considered. During ionization, part of the developing charge will be transferred to the α-hydrogen which should then *attract* rather than *repel* the leaving group (**39**) in a kind of hydrogen bonding interaction.[38] Tertiary systems could not enjoy this effect to the same extent.

39

[38]Recent calculations by Prof. W. L. Jorgensen support this possibility in a general sense. Compare **39** with the structure of the HCl-solvated isopropyl cation (*J. Amer. Chem. Soc.*, **99**, 280 (1977)).

Equilibrating Tertiary 2-Norbornyl Cations

9.1. Introduction

The 1,2-di-*p*-anisyl-2-norbornyl cation is a particularly fascinating species with special historic interest. It is the first 2-norbornyl cation to be clearly characterized as a classical ion. It is also the first member to be investigated of a highly informative group of tertiary 2-norbornyl cations capable of existing either as a symmetrical σ-bridged nonclassical ion or as a pair of rapidly equilibrating classical ions. The experimental results clearly established the 1,2-di-*p*-anisyl-2-norbornyl cation to be the latter. The discovery was made by the individual who is still one of the leading exponents of σ-bridged cations. It must have been a heart-rending experience.

9.2. Theoretical Considerations

The proposed σ-bridged structure (**1**) for the 2-norbornyl cation has been interpreted as a resonance hybrid (**1**) of three canonical structures (**1′**, **1″**, **1‴**).[1] It has been suggested that the last structure cannot contribute

$$ \tag{1} $$

significantly to the resonance hybrid.[2] Consequently, only canonical structures **1′** and **1″** need now be considered as significant contributors to the proposed resonance hybrid.

[1] S. Winstein and D. Trifan, *J. Amer. Chem. Soc.*, **74**, 1147, 1154 (1952).
[2] P. v. R. Schleyer, M. M. Donaldson, and W. E. Watts, *J. Amer. Chem. Soc.*, **87**, 375 (1965).

Under stable ion conditions the tertiary 2-methyl-2-norbornyl cation is reported to be some $7.5\,kcal\,mol^{-1}$ more stable than the secondary 2-norbornyl cation.[3] With such a large difference in the stability between secondary (**3**) and tertiary (**2**) canonical structures, significant resonance involving these structures would not be expected (2). Consequently, it was early suggested that tertiary 2-norbornyl cations, such as 2-methyl-2-norbornyl, should be essentially classical in nature.[4]

$$\Delta E \approx 5.5-7.5 \quad kcal\,mol^{-1} \tag{2}$$

Unfortunately, we do not now have really satisfactory data for the stabilities of cations under stable ion conditions.[5] In the absence of such data, we can utilize the relative stabilities of the transition states in solvolytic processes to estimate the relative stabilities of the intermediates.[6] The effect of various groups on the rate of solvolysis of 2-norbornyl chloride[7] is summarized in Figure 9.1. As an estimate, ignoring the effects of minor changes in ground-state energies, we can take $1.36 \log k/k_H$ as the stabilization of the transition state, with the stabilization of the free ion estimated to be modestly larger.[6] For example, taking $\log k/k_H$ as 5, we estimate the transition state for 2-methyl-2-norbornyl to be more stable than 2-norbornyl by $5 \times 1.36 = 6.8\,kcal\,mol^{-1}$. This compares with the experimental value of $7.5\,kcal\,mol^{-1}$ under stable ion conditions.[3]

Thus, if resonance is not to be a significant factor in the 2-methyl-2-norbornyl cation (2),[4] it should obviously be even less significant in the 2-phenyl- (**4**) and 2-*p*-anisyl-2-norbornyl (**6**) cations where the differences in energies of the secondary (**5, 7**) and tertiary ions (**4, 6**) are much larger (3, 4).

$$\Delta E \approx 12\,kcal\,mol^{-1} \tag{3}$$

[3] R. Hasaline, E. Huang, K. Ranganayakulu, T. S. Sorensen, and N. Wong, *Can. J. Chem.*, **53**, 1876 (1975); R. Hasaline, N. Wong, T. S. Sorensen, and A. J. Jones, *ibid.*, **53**, 1891 (1975); T. S. Sorensen, *Accounts Chem. Res.*, **9**, 257 (1976).

[4] C. A. Bunton, *Nucleophilic Substitution at a Saturated Carbon Atom*, Elsevier, New York, 1963.

[5] A detailed study to provide such information is now underway by E. M. Arnett and his coworkers.

[6] G. S. Hammond, *J. Amer. Chem. Soc.*, **77**, 334 (1955).

[7] H. C. Brown and K. Takeuchi, *J. Amer. Chem. Soc.*, **90**, 2691 (1968).

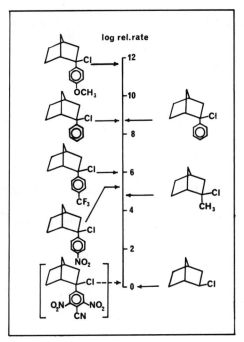

Figure 9.1. Relative rates of ethanolysis of 2-norbornyl derivatives (relative to *exo*-norbornyl chloride = 1.00).

$$\Delta E \approx 15 \text{ kcal mol}^{-1} \qquad (4)$$

6 OCH$_3$ **7** OCH$_3$

The question then arises as to what will happen if we introduce an identical substituent into the 1-position. Now the two canonical structures (**8′**) and (**8″**) become identical (5), just as are (**1′**, **1″**) in the parent system (1).

$$\equiv \qquad \longleftrightarrow \qquad (5)$$

R   R   R   R   R   R
8   **8′**   **8″**

Will we have a symmetrical σ-bridged nonclassical ion **8**, or will the species exist as a rapidly equilibrating pair (**9, 10**) of unsymmetrical classical cations (**6**)?

$$(6)$$

$$\text{R} \qquad \text{R} \qquad\qquad \text{R} \qquad \text{R}$$
$$\textbf{9} \qquad\qquad\qquad \textbf{10}$$

9.3. The 1,2-Di-p-anisyl-2-norbornyl Cation

The first objective test of the σ-bridged concept in the 2-norbornyl system appears to be the study by Schleyer, Kleinfelter, and Richey on the 1,2-di-*p*-anisyl-2-norbornyl cation.[8] The authors stated: "The unusual properties of carbonium ions in the bicyclo[2.2.1]heptane series have been widely interpreted to support a bridged structure. Although a number of stable substituted monoarylnorbornyl cations[9] have been studied, the structure of such carbonium ions—bridged or simple—remains unproved because of the difficulty of obtaining unambiguous structural information. We believe that the symmetry properties of diarylnorbornyl carbonium ions (derived from 1,2-diarylnorbornanols) make definitive determination of their structure possible."

The authors examined a number of characteristics.

A. Ultraviolet Spectral Behavior

The ion was generated in concentrated sulfuric acid. The uv spectrum was similar to that of the 2-*p*-anisyl-2-norbornyl cation (**6**) without the changes anticipated for extended conjugation in a symmetrical σ-bridged cation (**13**).

$$(7)$$

$$\text{OCH}_3 \qquad\qquad \text{OCH}_3 \qquad\qquad \text{OCH}_3 \qquad\qquad \text{OCH}_3 \qquad \text{OCH}_3 \qquad\qquad \text{OCH}_3$$
$$\textbf{11} \qquad\qquad\qquad\qquad \textbf{12} \qquad\qquad\qquad\qquad \textbf{13}$$

[8]P. v. R. Schleyer, D. C. Kleinfelter, and H. G. Richey, Jr., *J. Amer. Chem. Soc.*, **85**, 479 (1963).
[9]P. D. Bartlett, E. R. Webster, C. E. Dills, and H. G. Richey, Jr., *Ann.*, **623**, 217 (1959); P. D. Bartlett, C. E. Dills, and H. G. Richey, Jr., *J. Amer. Chem. Soc.*, **82**, 5414 (1960).

B. Thermodynamic Stability Measurements

Ion **6** was half formed from the carbinol in 41% sulfuric acid. If the stability of the ion is enhanced by the second *p*-anisyl group, as in **13**, the ion should be half formed in more dilute sulfuric acid. However, it actually required 51% sulfuric acid, in accord with **11** ⇌ **12**, where the inductive effect of the 1-*p*-anisyl group should destabilize the cation (**6**).

C. Chemical Behavior

Ion **6** does not undergo bromination in 4.5 M H_2SO_4 in CF_3CO_2H. The bromination of tri-*p*-anisylmethyl cation is relatively slow (~3 hours). But the 1,2-di-*p*-anisyl-2-norbornyl cation took up the bromine in less than 1 min. This corresponds to the presence of a *p*-anisyl group not conjugated with the positive charge (**11** ⇌ **12**).

D. Nuclear Magnetic Resonance

The nmr spectrum indicated equivalence of both aryl rings (anisyl quartet and methoxyl singlet) and symmetry in the remainder of the molecule, compatible either with the symmetrical structure **13** or with rapidly equilibrating unsymmetrical structures **11** ⇌ **12**. However, on cooling to −70°C, changes in the spectrum were observed which indicated impending nonequivalence of the two aryl rings.

The authors reached the following conclusion:[8] "We conclude on the basis of all four experimental criteria that diarylnorbornyl carbonium ions possess rapidly equilibrating asymmetric ion structures [**11** ⇌ **12**]. This result does not rule out the possibility of bridged structures for nonaryl substituted norbornyl or other nonclassical carbonium ions, but it stresses the desirability of reopening the question of the structure of such intermediates."

It has been argued that a possible difficulty with this experiment is the fact that the two *p*-anisyl groups cannot be coplanar for steric reasons.[10,11] However, such steric difficulties do not prevent the three *p*-anisyl groups in the tri-*p*-anisylmethyl cation from stabilizing the system (pK_R for *p*-An$_3$C$^+$ 0.82, *p*-An$_2$CH$^+$ −1.24), even though the steric effects appear even more serious in the trityl system.[12] In the next section we shall discuss a system where such steric effects should not be a factor, but rapid equilibration is observed.

[10]P. D. Bartlett, *Nonclassical Ions*, Benjamin, New York, 1965, p. 474.

[11]G. D. Sargent, *Carbonium Ions*, Vol. III, G. A. Olah and P. v. R. Schleyer, Eds., Wiley–Interscience, New York, 1972, pp. 1132–1136.

[12]N. C. Deno and A. Schriesheim, *J. Amer. Chem. Soc.*, **77**, 3051 (1955).

9.4. The 1,2-Diphenyl-2-norbornyl Cation

The introduction of a phenyl group into the 2-position of *exo*-norbornyl chloride results in a rate enhancement in solvolysis approaching 10^9 (Figure 9.1). Consequently, if a second phenyl group at the 1-position could stabilize the system by making resonance possible (6), such a substituent would be expected to enhance the rate.

However, we observed that the rate for 1,2-diphenyl-*exo*-norbornyl chloride (**15**) is not faster but 16 times slower than 2-phenyl-*exo*-norbornyl chloride (**14**).[13] The decrease in rate is the anticipated inductive effect of a

	14	**15**
RR (25°C):	1.00	1/16

1-phenyl substituent which is not stabilizing a developing electron deficiency at C1 (8).

$$(8)$$

Examination of the 1,2-diphenyl-2-norbornyl cation by nmr has confirmed the conclusion that the system is best described as a pair of rapidly equilibrating classical cations (**16** ⇌ **17**).[14]

Finally, the phenanthrene derivative **18** dissolves in fluorosulfonic acid to give the classical equilibrating cations (**19** ⇌ **20**), recognizable by a uv

$$(9)$$

[13] Unpublished research with M.-H. Rei.
[14] G. A. Olah and G. Liang, *J. Amer. Chem. Soc.*, **96**, 195 (1974).

spectrum very similar to that of 9-protonated phenanthrene and by the broadening of all proton signals in the nmr other than that of the bridgehead proton at $-110°C$ (9).[15] Steric effects, such as have been postulated to account for the behavior of the 1,2-di-*p*-anisyl-2-norbornyl cation,[10,11] do not appear to be a factor in this system.

9.5. The 1,2-Dimethyl-2-norbornyl System

The steric argument is not applicable to the 1,2-dimethyl-2-norbornyl system.[16,17] Here the introduction of a methyl group in the 1-position of the *endo* isomer (**21**) increases the rate of solvolysis in 80% acetone[17] over that shown by the parent system (**22**) by a factor of 8.6 (10). Presumably, the

$$\text{(10)}$$

RR (25°C):   8.6   1.00

increase is the result of the combined steric and inductive effect of the 1-methyl substituent.

The question is as to the effect of a 1-methyl substituent in the *exo* isomer (**23**). If the 1-methyl group were to cause resonance to return (2, 5) to the parent structure (**24**), its effect should be far greater. However, the effect is the same (11).

$$\text{(11)}$$

RR (25°C):   8.5   1.00

In spite of the failure to detect any resonance in the system, 1,2-dimethyl-2-norbornyl *p*-nitrobenzoate exhibits an *exo*:*endo* rate ratio at 25°C of 564. The value becomes 875 corrected for internal return.[18] (This compares with an *exo*:*endo* rate ratio of 885 for the parent system, **24**:**22**.[19]) The product is ≥99.7% *exo*, ≤0.3% *endo*.[16] The data provide a

[15]R. M. Cooper, M. C. Grossel, and M. J. Perkins, *J. Chem. Soc. Perkin II*, 594 (1972).
[16]H. C. Brown and M.-H. Rei, *J. Amer. Chem. Soc.*, **90**, 6216 (1968).
[17]Unpublished research with M. Ravindranathan.
[18]H. Goering and K. Humski, *J. Amer. Chem. Soc.*, **90**, 6213 (1968).
[19]S. Ikegami, D. L. Vander Jagt, and H. C. Brown, *J. Amer. Chem. Soc.*, **90**, 7124 (1968).

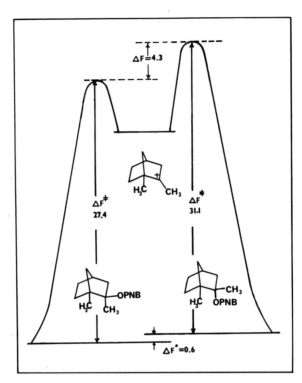

Figure 9.2. Free-energy diagram for the solvolysis of the 1,2-dimethyl-2-norbornyl *p*-nitro-benzoates in 80% acetone at 25°C.

free-energy diagram (Figure 9.2) that is again remarkably similar to those for the parent 2-norbornyl system (Figures 6.3. and 6.4). Again the question arises as to whether the difference in the energy of the two transition states arises from nonclassical stabilization of the *exo*, or from steric destabilization of the *endo*. In this case it is possible to give an unambiguous answer. The similarity in the *exo* (11) and the *endo* (10) isomers can only mean that there is no significant charge delocalization from the 2- to the 1-position. Without such charge delocalization, it is pointless to attribute the lower energy of the *exo* transition state to its stabilization by σ-bridging.

In this system it is further possible to establish that the intermediate formed is not the symmetrical σ-bridged nonclassical ion **25**.

Solvolysis of optically active 1,2-dimethyl-*exo*-norbornyl *p*-nitro-benzoate in 90% aqueous acetone gives alcohol with 9% retention.[18] Similarly, methanolysis of optically active 1,2-dimethyl-*exo*-norbornyl chloride gives methyl ether with 14% retention.[20] Goering and Clevenger conclude that they are trapping a rapidly equilibrating classical ion or ion pair before it has achieved complete equilibration (12).

$$(12)$$

Even under stable ion conditions the 1,2-dimethyl-2-norbornyl cation exists as a rapidly equilibrating pair of classical cations.[21,22] For example, Sorenson observed that in the 1,2,3,3- and 1,2,7,7-tetramethyl-2-norbornyl cations changes were observed with temperature that indicated that the system was a rapidly equilibrating pair of classical cations.[22] Yet the equilibration was not frozen out even at temperatures as low as −122°C, revealing how fast such Wagner–Meerwein shifts can take place (Chapter 13).

From Goering's results it appears that the rate of equilibration of the 1,2-dimethyl-2-norbornyl cation must be approximately 10 times the rate of solvent capture. No one has yet described a successful capture of the unsymmetrical 2-norbornyl cation or ion pair produced in a solvolytic process. (However, as will be described later, in Chapter 12, unsymmetrical 2-norbornyl cations or ion pairs have been successfully captured in a number of other processes.) Does this simple difference in the two systems make it desirable to ascribe the very similar free-energy diagrams for 1,2-dimethyl-2-norbornyl (Figure 9.2) and 2-norbornyl (Figures 6.3. and 6.4) to totally different physical causes?

[20]H. L. Goering and J. V. Clevenger, *J. Amer. Chem. Soc.*, **94**, 1010 (1972).
[21]G. A. Olah, J. R. De Member, C. Y. Lui, and R. D. Porter, *J. Amer. Chem. Soc.*, **93**, 1442 (1971).
[22]T. S. Sorensen and K. Ranganayakulu, *Tetrahedron Lett.*, 2447 (1972).

9.6. Conclusion

These tertiary 2-norbornyl derivatives all behave in a simple, consistent manner. It appears clear that the higher rates of the *exo* isomers and the lower energies of the *exo* transition states (Figure 9.2) cannot be the result of σ-bridging. The only other explanation that has received detailed consideration is steric hindrance to ionization in the *endo* isomers of these rigid bicyclic systems (Chapter 8), causing decrease in the rate of the *endo* isomer and an increase in the energy of the *endo* transition state (Figure 9.2). The problem is how to extrapolate from these well-behaved tertiary derivatives to the parent secondary structure. Application of the tool of increasing electron demand for this purpose is described in the next chapter.

Comments

The direct observation of an equilibrating set of 1,2-disubstituted-2-norbornyl cations (**9**, **10**) rigorously excludes a symmetrical, σ-bridged structure (**8**) but the possibility of a set of rapidly equilibrating unsymmetrical partially bridged ions remains. On the basis of ^{13}C and ^{1}H chemical shift deviations from additivity, Olah has suggested the latter for the 1,2-dimethyl-2-norbornyl cation.[21] Although this conclusion is not accepted by Brown, there is nothing wrong with partially bridged ions from a theoretical viewpoint. There should be a continuum from hyperconjugation without significant motion toward bridging, to unsymmetrical bridging,[23] to symmetrical bridging. Hyperconjugation and bridging are related rather than independent phenomena.[24] Partial bridging, if present in **9,10** (R = CH$_3$), probably contributes little to the stabilization of the species (and to stabilization of the 2-methyl-2-norbornyl cation).

Sorensen's conclusion[3] that the tertiary 2-methyl-2-norbornyl cation is only 7.5 kcal/mole more stable than the 2-norbornyl cation in superacid is very significant since it indicates that the secondary species enjoys about 7 kcal/mole of special stabilization. The typical value for tertiary-secondary energy differences, upon which this estimate is based, is much higher; 14.5 ± 0.5 kcal/mole has been determined directly for the 2-butyl $\rightarrow$ t-butyl cation isomerization in super acid.[25] Very similar results have been obtained in the gas phase by Solomon and Field,[26] who conclude "that the 2-norbornyl cation has 10 kcal/mole more stability than that which would be expected for a secondary species."

The special stabilization of the 2-norbornyl cation found experimentally in the gas phase, in superacid solution, and in solvolysis, constitutes especially compelling evidence consistent with the bridged, nonclassical, but not with the rapidly equilibrating classical structural formulation for this species. (For further elaboration, see Comments, Chapter 13.)

---

[23]STO-3G calculations indicate the 1-propyl cation to be partially bridged with an 83° C—C—C angle. While this structure no longer is a local minimum at more refined ab initio levels, the principle involved is indicated. See L. Radom, J. A. Pople, V. Buss, and P. v. R. Schleyer, *J. Amer. Chem. Soc.*, **94**, 311 (1972), and P. C. Hariharan, L. Radom, J. A. Pople, and P. v. R. Schleyer, *ibid.*, **96**, 599 (1974).

[24]G. A. Olah and G. Liang, *J. Amer. Chem. Soc.*, **95**, 3792 (1973).

[25]E. W. Bittner, E. M. Arnett, and M. Saunders, *J. Amer. Chem. Soc.*, **98**, 3734 (1976).

[26]J. J. Solomon and F. H. Field, *J. Amer. Chem. Soc.*, **98**, 1567 (1976); cf. R. D. Weiting, R. H. Staley, and J. L. Beauchamp, *ibid.*, **96**, 7552 (1974), and F. Kaplan, P. Cross, and R. Prinstein, *ibid.*, **92**, 1445 (1970).

10
Effect of Increasing Electron Demand

10.1. Introduction

The nonclassical ion theory is a "soft" theory, exceedingly flexible. In Section 1.8 several examples were presented to show how conflicting data and interpretations could be used and, in the past, have on occasion been so used to support the nonclassical position in spite of the evident contradictions. What is needed is an objective measure of π- and σ-participation. The Gassman–Fentiman approach[1] (Section 7.2) offered promise of such an objective measure. Consequently, we undertook to apply it to a large number of representative systems involving possible π- and σ-participation. The results, to be described in this chapter, proved both promising and consistent. However, applied to 2-norbornyl it failed to detect σ-participation.

10.2. Basic Considerations

A brief review may be in order. As was pointed out earlier (Chapter 7), the position that the importance of neighboring-group participation should diminish as an incipient cationic center is stabilized by structure modification[2] has been generally accepted.[3] It is now supported by a great deal of data.[4] However, the original study by Gassman and Fentiman on the *anti*-7-norbornenyl system is the most satisfying.[1]

---

[1]P. G. Gassman and A. F. Fentiman, Jr., *J. Amer. Chem. Soc.*, **92**, 2549 (1970).
[2]S. Winstein, B. K. Morse, E. Grunwald, K. C. Schreiber, and J. Corse, *J. Amer. Chem. Soc.*, **74**, 1113 (1952).
[3]P. D. Bartlett, *Nonclassical Ions*, Benjamin, New York, 1965.
[4]W. J. le Noble, *Highlights of Organic Chemistry*, M. Dekker, New York, 1974.

Thus they observed that the rate enhancement of 10^{11} observed in the parent secondary derivatives[5] decreases with the introduction of stabilizing groups at the 7-position (**2**) and effectively vanishes with the *p*-anisyl group.

$\rho+=-5.27$ **1**

$\rho+=-2.30$ **2**

Z =	**1**	**2**
7-H	1.00	10^{11}
3,5-$(CF_3)_2$	1.00	255,000
p-CF_3	1.00	34,000
p-H	1.00	41.5
p-CH_3O	1.00	3.4

The reaction constant $\rho+$, from the relationship, $\log(k/k_H) = \rho+\sigma+$,[6] provides a convenient measure of the effect. Thus $\rho+$ for **1** is -5.27; $\rho+$ for **2** is -2.30. With a substituent even more stabilizing than *p*-OCH_3, namely, *p*-$N(CH_3)_2$, the behavior of **2** parallels **1** (Figure 10.1).

It is desirable to distinguish between π- and σ-participation and π- and σ-conjugation.

In the cyclopent-3-en-1-yl derivative (**3**), participation of the double bond, if it were to occur, would lead to a less negative value of $\rho+$. This

3   **4**

would be π-participation. On the other hand, in **4** the cationic center would be stabilized by allylic conjugation with the cationic center. This should lead to an increased (less negative) value for $\rho+$. We do not term this π-participation, but π-conjugation.

This is nicely illustrated by the $\rho+$ values for **5** and **6**.[7]

$\rho+=$ -4.60 -2.52

[5]S. Winstein, M. Shatavsky, C. Norton, and R. B. Woodward, *J. Amer. Chem. Soc.*, **77**, 4183 (1955).

[6]H. C. Brown and Y. Okamoto, *J. Org. Chem.*, **22**, 485 (1957).

[7]H. C. Brown, M. Ravindranathan, and M. M. Rho, *J. Amer. Chem. Soc.*, **98**, 4216 (1976).

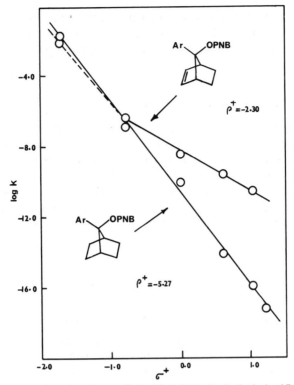

Figure 10.1. Effect of increasing electron demand on the rates of solvolysis of 7-arylnorbornyl and 7-aryl-*anti*-*norbornenyl p*-nitrobenzoates.

Unfortunately, in the literature there has not been the same care in differentiating between σ-participation and σ-conjugation (hyperconjugation). We shall attempt to avoid this ambiguity by using the term "σ-participation" for a direct interaction through space of a *p*-orbital with the electrons in a carbon–carbon bond, as indicated in the diagrams from Streitwieser[8] (Figures 5.1 and 5.2). On the other hand, interaction of the carbonium ion center in the cyclopropylcarbinyl cation in the bisected arrangement does not involve such a σ-bridge through space, but sideways interaction of orbitals, and is better termed "σ-conjugation."

In applying the tool of increasing electron demand to the cyclopropyl-carbinyl system (Section 10.4), we will be in position to determine from the values of $\rho+$ whether we are observing σ-electronic contributions. For a decision as to whether these are σ-participation or σ-conjugation we shall have to decide as to whether the electronic contributions are made in the parallel conformation (Figures 5.1 and 5.2), or in the bisected conformation.

[8] A. Streitwieser, Jr., *Solvolytic Displacement Reactions*, McGraw-Hill, New York, 1962.

We would be faced with a similar problem in the 2-norbornyl system (Section 10.5), except that the studies fail to reveal any evidence for significant σ-electronic contributions that are greater in the *exo* derivative than in the *endo*.

10.3. Electron Demand in π-Systems

The parent secondary 7-norbornenyl system is one where the evidence for π-participation is unambiguous.[5] The tool of increasing electron demand yields $\rho+ -2.30$ for **2** as contrasted to the value of -5.27 for **1**, completely consistent with the earlier conclusion.

The parent secondary cyclopent-3-en-1-yl system is one where π-participation appears to be absent.[9] Application of the tool of increasing electron demand confirms that conclusion (**7** and **8**).[10]

	7	8
$\rho+ =$	-3.82	-3.92

In systems involving the 2-norbornyl species, it is customary to compare the *exo* isomer with the *endo*, where it is postulated that neither π- nor σ-contributions are significant.[11] In the 2-norbornenyl system[12] both the values of $\rho+$ and the *exo*:*endo* rate ratio fail to reveal π-participation over the first three members (**9, 10**). Only with the larger demand of the 3,5-$(CF_3)_2C_6H_3$ derivative and the secondary derivative do we observe small increases in the *exo*:*endo* rate ratio attributable to π-participation.[12] (For the large effects of even larger electron demand, see Section 10.6.)

$\rho+ = -4.17$

$\rho+ = -4.21$

Z =	9	10
p-OCH$_3$	1.00	312
p-H	1.00	201
p-CF$_3$	1.00	283
3,5-(CF$_3$)$_2$	1.00	447
2-H	1.00	7000[13]

[9]S. Winstein and J. Sonnenberg, *J. Amer. Chem. Soc.*, **83**, 3235 (1961).
[10]E. N. Peters and H. C. Brown, *J. Amer. Chem. Soc.*, **97**, 7454 (1975).
[11]S. Winstein and D. Trifan, *J. Amer. Chem. Soc.*, **74**, 1154 (1952).
[12]H. C. Brown and E. N. Peters, *J. Amer. Chem. Soc.*, **97**, 7442 (1975).
[13]The value is for the acetolysis of the brosylate. S. Winstein, H. Walborsky, and K. C. Schreiber, *J. Amer. Chem. Soc.*, **72**, 5795 (1950).

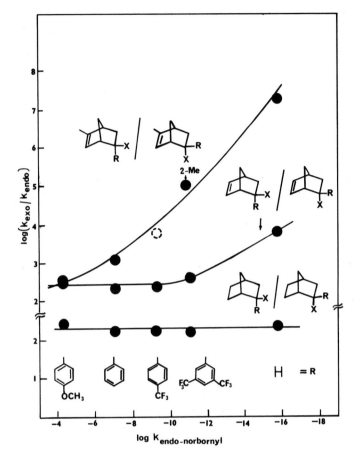

Figure 10.2. Effect of increasing electron demand on the *exo*:*endo* rate ratio in 2-norbornyl, 2-norbornenyl, and 5-methyl-2-norbornenyl derivatives.

The results are summarized in Figure 10.2.

We conclude that the essentially constant factor of 300 over the range where it exhibits no sensitivity to electron demand must result from steric hindrance to ionization (Chapter 8). If it is possible to carry this factor over to the secondary derivative, the *exo*:*endo* rate ratio of 7000 would be made up of a factor of approximately 300 resulting from steric hindrance to ionization and a factor of approximately 24 attributed to π-participation.

The solvolysis products correspond to this interpretation. Thus, both **9** and **10** (*p*-CH₃O) give the open *exo*-alcohol **11**, whereas 2-H and 3,5-(CF₃)₂ give predominantly the nortricyclyl derivative **12**.

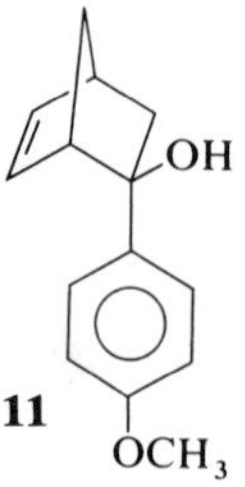

11

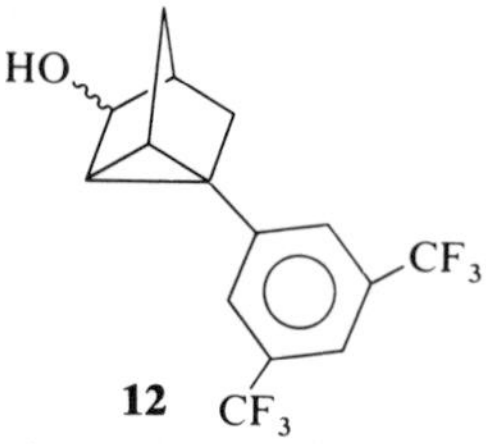

12

Activation of the double bond by a methyl group in the 5-position greatly increases π-participation (**13, 14**).[14] Only with the most activating group examined, p-OCH$_3$, is the π-participation essentially absent, with an *exo*:*endo* rate ratio corresponding to that observed in the parent system (**9, 10**).

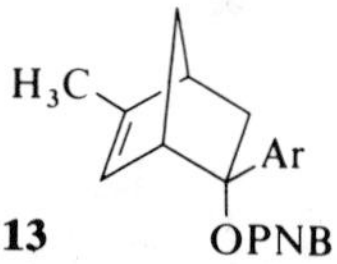

13 $\rho+ = -4.19$ 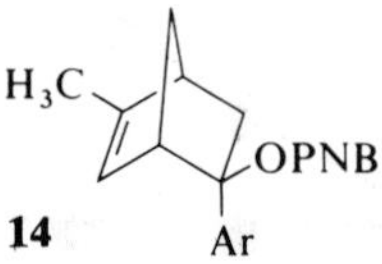   **14** $\rho+ = -3.27$

Z =	**13**	**14**
p-OCH$_3$	1.00	354
p-H	1.00	1,260
p-CF$_3$	1.00	6,700
3,5-(CF$_3$)$_2$	1.00	17,800
2-H	1.00	22,000,000[15]

The *exo*:*endo* rate ratio in the 2-aryl-2-benzonorbornenyl derivatives is higher, ~3000, and does not vary with electron demand (**15, 16**).[16]

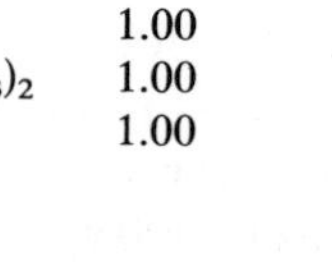

15 $\rho+ = -4.51$ 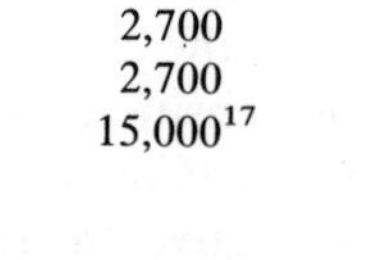   **16** $\rho+ = -4.50$

Z =	**15**	**16**
p-OCH$_3$	1.00	3,300
p-H	1.00	2,850
p-CF$_3$	1.00	2,700
3,5-(CF$_3$)$_2$	1.00	2,700
2-H	1.00	15,000[17]

[14]H. C. Brown, E. N. Peters, and M. Ravindranathan, *J. Amer. Chem. Soc.*, **97**, 7449 (1975).
[15]Private communication, C. F. Wilcox.
[16]H. C. Brown, S. Ikegami, K.-T. Liu, and G. L. Tritle, *J. Amer. Chem. Soc.*, **98**, 2531 (1976).
[17]The value is for the acetolysis of the brosylate. H. C. Brown and G. L. Tritle, *J. Amer. Chem. Soc.*, **90**, 2689 (1968). Corrected for internal return the value is 62,000; see J. Dirlam, A. Diaz, S. Winstein, W. P. Giddings, and G. C. Hanson, *Tetrahedron Lett.*, 3133 (1969).

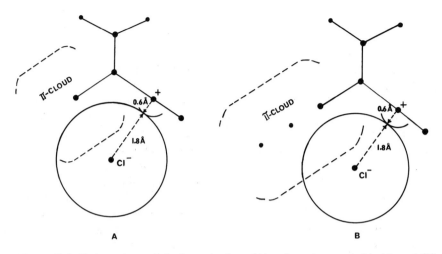

Figure 10.3. End-on views of the ion pairs from (A) *endo*-norbornenyl chloride and (B) *endo*-benzonorbornenyl chloride.

Again, the insensitivity of the *exo*: *endo* rate ratio to increasing electron demand indicates that aryl participation cannot be a significant factor in the high *exo*: *endo* rate ratio exhibited by these tertiary derivatives. Why should the *exo*: *endo* rate ratio be so much higher than in the related 2-aryl-2-norbornenyl derivatives (**9, 10**)? An examination of Figure 10.3 reveals a reasonable explanation. The U-shaped cavity of the molecule is considerably accentuated in the benzonorbornenyl derivative.

Activation of the benzo moiety by the introduction of a methoxy group in the *homopara* position does result in increased *exo*: *endo* rate ratios which increase with increasing electron demand (**17, 18**).[18] A puzzle is the significant increase in p^+ (to -4.10) for the *endo* derivatives (**17**).

Z =	**17**	**18**
p-OCH$_3$	1.00	7,000
p-H	1.00	14,500
p-CF$_3$	1.00	25,000
3,5-(CF$_3$)$_2$	1.00	34,600
2-H	1.00	830,000[19]

17 $p+ = -4.10$ **18** $p+ = -3.71$

[18]H. C. Brown and K.-T. Liu, *J. Amer. Chem. Soc.*, **91**, 5909 (1969).
[19]The value is for the acetolysis of the brosylate. D. V. Braddon, G. A. Wiley, J. Dirlam, and S. Winstein, *J. Amer. Chem. Soc.*, **90**, 1901 (1968).

Consequently, in the methoxy derivatives the *exo*: *endo* rate ratios must be made up of components attributable to aryl participation and components attributed to steric hindrance to ionization.

There is a major unresolved problem with the secondary derivatives. As the benzo moiety is deactivated by the introduction of substituents, such as nitro, aryl participation would be expected to decrease. We should then anticipate approaching a limiting value for the *exo*: *endo* rate ratio attributable to steric hindrance to ionization. However, this is not observed.[20] There is some evidence that the presence of deactivating groups transforms the solvolysis process from one involving ionization of the *exo* and *endo* isomers to produce a common intermediate to a reaction in which solvent participation occurs and no common intermediate is formed.[21] We shall discuss this further in Chapter 11 when we examine the behavior of 2-norbornyl derivatives containing deactivating substituents. Here also there is a decrease in *exo*: *endo* rate ratios with considerably different products realized from *exo* and *endo* derivatives.

10.4. Electron Demand in Cyclopropylcarbinyl Systems

A comparison of the effect of electron demand on the methylarylcyclopropylcarbinyl *p*-nitrobenzoates (20) with the related isopropyl derivatives (19) reveals major increases in electron supply with increasing electron demand.[22]

	19 OPNB $\rho+ = -4.76$	**20** OPNB $\rho+ = -2.78$
Z = *p*-OCH$_3$	1.00	505
p-H	1.00	25,300
p-CF$_3$	1.00	285,000
3,5-(CF$_3$)$_2$	1.00	1,210,000

It is clear that the data reveal major σ-electronic contributions from the cyclopropyl ring to the developing carbonium ion center. However, as discussed earlier, there is a difficulty in deciding whether these σ-electronic contributions should be classified as σ-participation or as σ-conjugation.

The 3-aryl-3-nortricyclyl derivatives provide a means of resolving this ambiguity. The geometrical requirements of the 3-nortricyclyl cation make

[20] H. Tanida, H. Ishitobi, T. Irie, and T. Tsushima, *J. Amer. Chem. Soc.*, **91**. 4512 (1969); H. Tanida, T. Irie, and T. Tsushima, *J. Amer. Chem. Soc.*, **92**, 3404 (1970).

[21] Unpublished research with C. J. Kim.

[22] E. N. Peters and H. C. Brown, *J. Amer. Chem. Soc.*, **95**, 2397 (1973).

σ-participation impossible (**21**).[23] Consequently, if such σ-participation

Ar

21

were the significant factor in **20** (Figure 5.1), the effect of increasing electron demand should be much less in the 3-arylnortricyclyl derivatives (**23**).

On the other hand, the bisected conformation, corresponding to σ-conjugation, is ideal in **21**. If this is the mechanism for σ-electron supply in **20** and similar cyclopropylcarbinyl derivatives, then the effect of electron demand in **23** should parallel that observed in **20**.[24]

Ar OPNB   Ar OPNB

22 $\rho+ = -5.27$ **23** $\rho+ = -3.27$

$Z =$	**22**	**23**
p-CH$_3$O	1.00	7,000
p-H	1.00	25,000
p-CF$_3$	1.00	3,010,000
3,5-(CF$_3$)$_2$	1.00	10,400,000
2-H	1.00	10^9 [25]

It should be observed that $\Delta\rho+$ is 1.98 for the system **20–19**, and 2.00 for the system **23–22**. Consequently, the tool of increasing electron demand can detect not only π-electron contributions, as previously discussed, but σ-electron contributions in the cyclopropylcarbinyl derivatives. However, the data clearly support the conclusion that in this system the σ-electron contributions must be classified as σ-conjugation, not σ-participation.

The introduction of electron withdrawing (**24**) or electron supplying (**25**) substituents into the cyclopropane ring results in predictable changes in $\rho+$ (**24, 25**).[26] Thus the presence of the two chlorine substituents reduce the electronic contributions from the cyclopropane ring to a value below that for the model isopropyl group (**19**).

Cl   CH$_3$

Cl—⟨⟩—Ar   H$_3$C—⟨⟩—Ar

24   OPNB   **25**   OPNB

$\rho+ = -4.99$ $\rho+ = -2.06$

[23] J. D. Roberts, W. Bennett, and R. Armstrong, *J. Amer. Chem. Soc.*, **72**, 3329 (1950).

[24] H. C. Brown and E. N. Peters, *J. Amer. Chem. Soc.*, **97**, 1927 (1975).

[25] The value is for the acetolysis of the brosylate. H. G. Richey, Jr., and N. C. Buckley, *J. Amer. Chem. Soc.*, **85**, 3057 (1963).

[26] H. C. Brown, E. N. Peters, and M. Ravindranathan, *J. Amer. Chem. Soc.*, **99**, 505 (1977).

10.5. Electron Demand in 2-Norbornyl Systems

Of course, the objective of this study was the 2-norbornyl system. Would the tool of increasing electron demand reveal σ-electronic contributions from the norbornyl system in the manner it had detected σ-electronic contributions from the cyclopropyl ring in cyclopropylcarbinyl derivatives[22] and π-electronic contributions in the 5-methyl-2-norbornenyl system?[14]

However, the data reveal an essentially constant *exo*:*endo* rate ratio (Figure 10.2) which does not change significantly with increasing electron demand (**26, 27**).[27]

	26	OPNB	$\rho+=-3.72$		27	Ar	$\rho+=-3.82$

$Z = p\text{-}CH_3O$	1.00	284
$p\text{-}CH_3$	1.00	232
$p\text{-}H$	1.00	127
$p\text{-}CF_3$	1.00	187
$3,5\text{-}(CF_3)_2$	1.00	176
H	1.00	280[28]

The similarity in the values of $\rho+$ for *exo* (**27**) and *endo* (**26**) should be noted, as well as the similarity to the value for cyclopentyl, ($\rho+=-3.82$, **7**). On the other hand, the $\rho+$ values for the isopropyl (**19**) and cyclohexyl (**5**) are more negative, -4.76 and -4.60, respectively. These values suggest that cyclopentyl, *exo*-norbornyl, and *endo*-norbornyl supply electrons to the electron-deficient center more effectively than do the isopropyl and cyclohexyl groups. Perhaps the strain in the former three systems make them better able to hyperconjugate and thereby stabilize the electron-deficient center. However, it is of the utmost importance for the nonclassical ion problem that *endo*-norbornyl is, if anything, slightly more effective than *exo*-norbornyl in thus stabilizing an electron-deficient center, as measured by $\rho+$.

Recently, Battiste has argued that the original proposal of Winstein's, that the rate of *exo* be compared with *endo*, be abandoned.[29] He therefore

[27] H. C. Brown, M. Ravindranathan, K. Takeuchi, and E. N. Peters, *J. Amer. Chem. Soc.*, **97**, 2900 (1975).

[28] For the acetolysis of the tosylate see P. v. R. Schleyer, M. M. Donaldson, and W. E. Watts, *J. Amer. Chem. Soc.*, **87**, 375 (1965). The brosylate shows *exo*:*endo* = 350. Corrected for internal return, this is 1600 (Section 6.3).

[29] M. A. Battiste and R. A. Fiato, *Tetrahedron Lett.*, 1255 (1975).

compares *exo*-norbornyl ($\rho+ = -3.82$) and *endo*-norbornyl ($\rho+ = -3.72$) with isopropyl ($\rho+ = -4.76$). He then concludes that both *exo*-norbornyl (the C1–C6 bond) and *endo*-norbornyl (the C1–C7 bond) are capable of supplying σ-electrons to the cationic center under the demand of the tertiary carbonium ion center. This again illustrates what a "soft" theory the nonclassical ion concept provides (Section 1.8). After 25 years, in which the *exo*:*endo* rate ratio was considered diagnostic for nonclassical ions, a supporter of the concept can seriously propose abandonment of this criterion and the adoption of a totally new reference.

The authors fail to make it clear why the cyclopentane system, where the carbonyl frequencies (ν_{CO}) in the ketones are similar, and the bond opposition forces in the parent systems are similar, is not a better model for *exo*- and *endo*-norbornyl.

I see no merit in this proposal in its present form and shall reserve judgment until the time when the authors demonstrate its utility in predicting new phenomena.

The tool of increasing electron demand has also been applied to 2-camphenilyl (**28, 29**).[30]

Z =	**28** OPNB	**29** Ar
p-OCH$_3$	1.00	44,300
p-H	1.00	48,400
p-CF$_3$	1.00	23,600
3,5-(CF$_3$)$_2$	1.00	26,900
2-H	1.00	1,240[31]

An examination of Figure 10.2 reveals the effect of increasing electron demand on the *exo*:*endo* rate ratios of three representative systems: (a) 5-methyl-2-norbornenyl, which reveals a major increase with demand; (b) 2-norbornyl, which reveals no change; and (c) 2-norbornenyl, which is evidently borderline, showing modest increases in *exo*:*endo* rate ratios only at the higher level of electron demand. (Much larger increases in the 2-norbornenyl system have been observed with still larger increase in electron demand, discussed in the next section.)

Pertinent data for these studies are summarized in Table 10.1.

---

[30] H. C. Brown and K. Takeuchi, *J. Amer. Chem. Soc.*, **90**, 5268 (1968).
[31] The value is for the acetolysis of the brosylate. A. Colter, E. C. Friedrich, N. J. Holness, and S. Winstein, *J. Amer. Chem. Soc.*, **87**, 378 (1965). See discussion in Section 7.4.

TABLE 10.1. Rate Constants (25°C) for Studies of the Effect of Increasing Electron Demand

System R—C—X	$10^6 k_1$, sec⁻¹, R = Ph, X = OPNB 80% acetone	ρ^+	$10^6 k_1$, sec⁻¹, R = Ph, X = Cl 90% acetone	ρ^+	$10^6 k_1$, sec⁻¹, R = Me, X = OPNB 80% acetone	$10^6 k_1$, sec⁻¹, R = H, X = OTs HOAc	$10^6 k_1$, sec⁻¹, R = H, X = OPNB 80% acetone[a]
Cyclobutyl	0.0018	−4.91[b]	27.4	−4.48[c]			1.27×10^{-11}
Cyclopentyl	2.6	−3.82[d]	7230	−4.10[c]	2.11×10^{-3e}	1.65[f]	5.17×10^{-11}
Cyclopent-3-enyl	0.798	−3.92[d]			1.41×10^{-4e}		6.25×10^{-12}
Cyclohexyl	0.0146	−4.60[g]	19.6	−4.65[c]	5.48×10^{-5h}	0.0488	1.59×10^{-12}
Cyclohex-2-enyl	2220	−2.52[g]					
7-Norbornyl	0.000102	−5.27[i]	0.0398	−5.64[c]		6.36×10^{-9j}	2.07×10^{-19}
anti-7-Norbornenyl	0.00423	−2.30[i]				904[j]	2.96×10^{-8}
exo-2-Norbornenyl	1.21	−4.21[d]			4.70×10^{-4d}	15.0[k,l]	4.91×10^{-10}
endo-2-Norbornenyl	0.0060	−4.17[d]			5.27×10^{-7d}	0.0019[k,l]	6.54×10^{-14}
exo-5-Methyl-2-norbornenyl	9.18	−3.27[m]			0.0572^m		1.43×10^{-4}
endo-5-Methyl-2-norbornenyl	0.0073	−4.19[m]			5.13×10^{-7m}		6.54×10^{-14}
exo-Benzonorbornenyl	0.278	−4.50[n]			3.59×10^{-5o}	2.49[l,n]	6.97×10^{-11}

Compound							
endo-Benzonorbornenyl	9.74×10^{-5}	-4.51[n]			5.5×10^{-9}[o]	1.7×10^{-4}[l,n]	4.77×10^{-15}
exo-6-Methoxy-benzonorbonenyl	1.36	-3.71[p]				310[l,q]	8.7×10^{-9}
endo-6-Methoxy-benzonorbornenyl	9.32×10^{-5}	-4.10[p]				0.00112[l,q]	1.05×10^{-14}
exo-Norbornyl	7.56	-3.82[r]			0.01[d]	23.3[s]	7.85×10^{-10}
endo-Norbornyl	0.0594	-3.72[r]			1.13×10^{-5}[d]	0.0828[s]	2.66×10^{-12}
exo-Camphenilyl	22.9	-3.65[t]			0.104[w]	10.96[v]	3.07×10^{-10}
endo-Camphenilyl	4.73×10^{-4}	-3.47[t]			2.31×10^{-5}[u]	0.0088[v]	2.49×10^{-13}
3-Nortricyclyl	0.365	-3.27[w]			0.00181	4.00[x]	1.26×10^{-10}
2-Adamantyl			459	-4.83[c]	1.43×10^{-4}	0.0059	1.92×10^{-13}
3-Methyl-2-butyl	0.00951	-4.76[y]			0.00022		
Cyclopropyl-methyl	241	-2.78[y]			37.5[z]		
Dimethylcyclo-propylmethyl	2990	-2.06[y]			917		
Dichlorocyclo-propylmethyl	0.0179	-4.99[y]			0.00022		

[a] Extrapolated values: (Reference 40). [b] M. Ravindranathan and C. Gundu Rao, research in progress. [c] H. Tanida and T. Tsushima, *J. Amer. Chem. Soc.*, **92**, 3397 (1970). [d] Reference 10. [e] Reference 12. [f] D. D. Roberts, *J. Org. Chem.*, **33**, 118 (1968). [g] Reference 7. [h] E. N. Peters and H. C. Brown, *J. Amer. Chem. Soc.*, **97**, 2892 (1975). [i] Reference 1. [j] Reference 5. [k] S. Winstein and M. Shatavsky, *J. Amer. Chem. Soc.*, **78**, 592 (1956). [l] Calculated from the ratio $k_{OBs}/k_{OTs} = 3$. [m] Reference 14. [n] Reference 16. [o] H. C. Brown and G. L. Tritle, *J. Amer. Chem. Soc.*, **88**, 1320 (1966). [p] Reference 18. [q] Reference 19. [r] Reference 27. [s] Reference 28. [t] Reference 30. [u] K. Takeuchi, Ph.D. Thesis, Purdue University, 1969. [v] Reference 31. [w] Reference 24. [x] Reference 25. [y] Reference 26. [z] H. C. Brown and E. N. Peters, *J. Amer. Chem. Soc.*, **95**, 2400 (1973).

10.6. Increasing Electron Demand in Secondary 2-Norbornyl

One of the arguments against the trends revealed in Figure 10.2 is that the extrapolation from the tertiary derivatives examined [p-CH$_3$O, p-H, p-CF$_3$, and 3,5-(CF$_3$)$_2$] to the secondary derivatives is both a long one and involves major changes in the steric effects of the 2-substituent.[32,33] Consequently, it would be desirable to vary the electron demand at position 2 while maintaining the steric environment essentially constant. Would this result in an increase in the *exo*:*endo* rate ratio as the increased electron demand called upon the norbornenyl (π-participation) and the norbornyl system (σ-participation) for increasing electron supply?

In an elegant study, Lambert and Mark explored this question.[34] They observed that the introduction of a tosyl group in the 3-position of *endo*-norbornenyl (**30, 31**) and *endo*-norbornyl tosylate (**32, 33**) decreases the rate of acetolyses as compared to the parent systems by factors of approximately 100,000.

	30 OTs	**31** OTs	**32** OTs	**33** OTs
RR (25°C):	1.00	0.0000067	1.00	0.0000017

The question now is the effect the increased electron demand at C2 will have on the *exo* isomers. The authors observed a major increase in the *exo*:*endo* rate ratio for the norbornenyl system from 2900[35] to 140,000 (**34:31**).

	34	**31** OTs
RR (25°C):	140,000	1.00

Clearly, the results correspond to enhanced π-participation in (**34**), as compared to *exo*-norbornenyl tosylate.

---

[32]G. D. Sargent, *Carbonium Ions*, Vol. III, G. A. Olah and P. v. R. Schleyer, Eds., Wiley–Interscience, New York, 1972.

[33]Private communication, P. v. R. Schleyer.

[34]J. B. Lambert and H. W. Mark, *Tetrahedron Lett.*, 1765 (1976).

[35]This value is somewhat lower than the value of 7000–8000 customarily used. The authors point out that the earlier value is not based on a direct measurement, see footnote *a* in Table IV of S. Winstein and M. Shatavsky, *J. Amer. Chem. Soc.*, **78**, 592 (1956). However, we have observed a value of 7000 for the mesylates under conditions where internal return should not be a significant factor and that value has been used in this book.

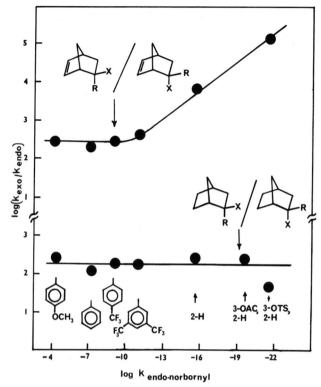

Figure 10.4. Effect of enhanced electron demand in the 2-norbornenyl and 2-norbornyl system on *exo* : *endo* rate ratio.

On the other hand, there is no corresponding increase in the *exo* : *endo* rate ratio for the 2-norbornyl derivatives (**35** : **33**).

RR (25°C):   48[36]   1.00

The markedly different effects of increasing electron demand on the two systems are revealed in Figure 10.4.

An alternative approach to increasing electron demand is the use of less nucleophilic solvents, such as trifluoroacetic acid. For example, this solvent

[36]The decrease in the *exo* : *endo* rate ratio may result from increased solvent participation in the *endo* isomer, as discussed in Chapter 11.

reveals enhanced aryl participation in the solvolysis of 1-phenyl-2-propyl tosylate (36:37).[37]

	CH$_2$CHCH$_3$	CH$_3$CHCH$_3$
	36 OTs	**37** OTs
RR (25°C):		
acetic acid	0.4	1.00
formic acid	0.58	1.00
trifluoroacetic acid	17.1	1.00

Consequently, if σ-participation is a factor in the 2-norbornyl system one would anticipate major increases in the *exo*:*endo* rate ratio as less nucleophilic solvents are used and the less solvated developing cationic center makes greater demand upon the system for electronic stabilization. However, the data do not support this. Thus, the *exo*:*endo* rate ratios for 2-norbornyl tosylates in acetic acid ($k_t = 280$, $k_\alpha = 1600$) show no major change in formic acid ($k_t \approx k_\alpha = 1600^{38}$) or in trifluoroacetic acid[39] ($k_t = 1120$).

10.7. Extrapolation of Data from the Tertiary to the Secondary Systems

An interesting alternative approach for extrapolating from the tertiary 2-aryl-2-norbornyl system to the secondary systems has been proposed by E. N. Peters.[40] He defines a linear free-energy relationship, $\log(k/k_0) = \rho\gamma+$, with group constants, $\gamma+$. These group constants are set equal to $\sigma+$ for the substituted 2-aryl substituents. By utilizing the data for 7-aryl-7-norbornyl *p*-nitrobenzoates,[1] he extrapolates the data to the secondary derivative to calculate a $\gamma+$ value for 7-H (Figure 10.5). He assumes that this value, γ_H^+ 2.53, will be a constant for hydrogen and utilizes it to extrapolate data from the tertiary derivatives to the secondaries.

[37]J. E. Nordlander and W. J. Kelly, *J. Amer. Chem. Soc.*, **91**, 996 (1969).

[38]Unpublished work with I. Rothberg.

[39]J. E. Nordlander, R. E. Gruetzmacher, W. J. Kelley, and S. P. Jindal, *J. Amer. Chem. Soc.*, **96**, 181 (1974). This is an exceptionally interesting paper to read. The authors evidently anticipated a large increase in the *exo*:*endo* ratio, similar to that observed for neighboring aryl participation (36:37). Failing to find such an increase, they did not draw the obvious conclusion, but instead developed an elaborate argument to retain their original position. The question of the importance of the k_s contribution in *endo*-norbornyl is discussed in the next chapter.

[40]E. N. Peters, *J. Amer. Chem. Soc.*, **98**, 5627 (1976).

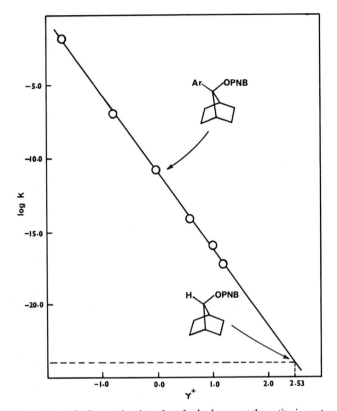

Figure 10.5. Determination of $\gamma+$ for hydrogen at the cationic center.

It is necessary to recalculate rates for secondaries, usually measured as the tosylates or brosylates in acetic acid, to the *p*-nitrobenzoates or chlorides used for the tertiary derivatives in the appropriate solvents. Over such large ranges of reactivities, considerable temperature extrapolations of the experimental data are also necessary. These extrapolations can result in considerable uncertainties in the data being treated. Yet the fits realized are remarkably good.

Consider, for example, the fit provided by a plot of the data for the 2-adamantyl[41] and 3-nortricyclyl derivatives (Figure 10.6).

In a case where participation begins early in the tertiary derivatives, with the *p*-methoxy derivative, both the 7-aryl-*anti*-norbornenyl derivatives involving participation and the 7-aryl-7-norbornyl derivatives not

[41]H. Tanida and T. Tsushima, *J. Amer. Chem. Soc.*, **92**, 3397 (1970).

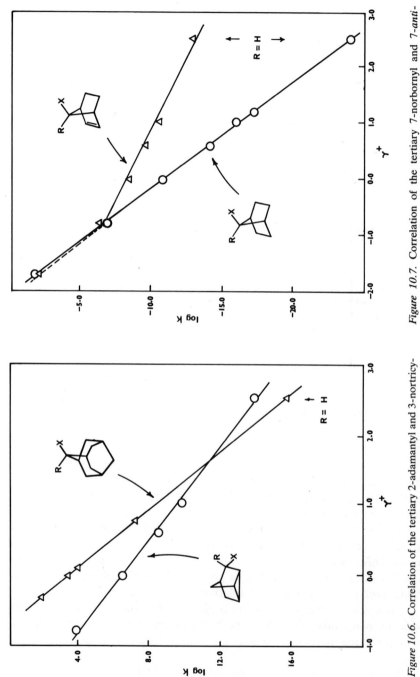

Figure 10.7. Correlation of the tertiary 7-norbornyl and 7-*anti*-norbornenyl derivatives with the secondary.

Figure 10.6. Correlation of the tertiary 2-adamantyl and 3-nortricyclyl derivatives with the secondary.

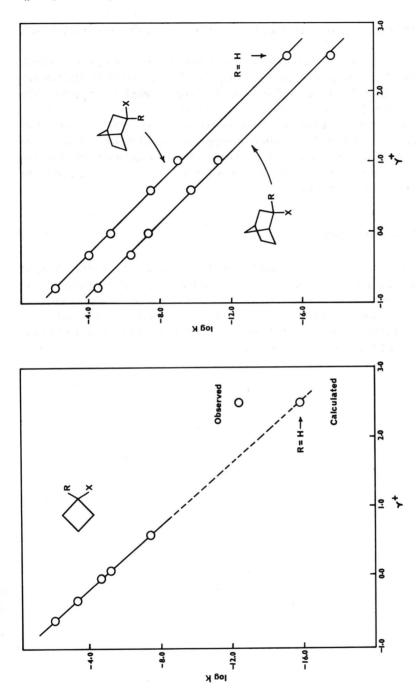

Figure 10.9. Correlation of the tertiary 2-norbornyl derivatives with the secondary.

Figure 10.8. Lack of correlation of the tertiary derivatives with the secondary in the cyclobutyl system.

involving participation correlate well with their respective secondary derivatives (Figure 10.7).

In a case where there is no σ-participation in the tertiary derivatives, but with the incursion of major participation in the secondary, the rate for the secondary should be much too fast to be so correlated. This appears to be the case for the cyclobutyl system[41] (Figure 10.8).

On the other hand, the 2-norbornyl derivatives reveal an excellent correlation[27] (Figure 10.9). There is simply no evidence for a break in the *exo* derivative corresponding to σ-participation, postulated to be absent in both *exo-* and *endo*-tertiary derivatives and in the *endo*-norbornyl, but present in *exo*-norbornyl![32,33]

10.8. Conclusion

The various approaches utilizing the tool of increasing electron demand at the cationic center have uniformly failed to reveal evidence for σ-participation in 2-norbornyl. What other approaches are available to use to test for the long postulated σ-participation in the solvolysis of 2-norbornyl, with delocalization of charge from the 2- to the 1- and 6-positions? In the next chapter we shall examine the use of substituents to test for such charge delocalization.

Comments

Classical carbocation theory is also "soft," as are, unfortunately, most theories of organic chemistry in their present stage of development. Quantitative before-the-fact predictions are seldom attempted, rather, theory is typically used after the fact to rationalize results.

The $\rho+$ values summarized in Table 10.1 afford an example. We have no way at present of predicting $\rho+$ to be expected of a given system before the experiments are run[42]; rather, the values obtained are interpreted afterwards. While it does not seem objectionable to compare closely related systems (e.g., **5** with **6**, **7** with **8**, **22** with **23**, and **24** with **25**), there are difficulties in achieving a more detailed understanding. There seems to be a general trend towards less negative $\rho+$ values with increasing reactivity, but exceptions are widespread. The phenyl derivatives of **16** and **23** have nearly the same rate constants, but $\Delta\rho+$ is 1.23 (out of a total range of only three units). The $\rho+$'s of **23** and **28** are almost identical, and yet the phenyl derivative of the latter is less reactive by 10^4.

Until we understand all of the factors involved,[42] we are not in a position to be able to interpret the significance of a given $\rho+$ value. The "tool of increasing electron demand" is only one possible measure of the "electron demand" of a system.

It is not the magnitude of $\rho+$, but rather the *break* in a Hammett–Brown plot (e.g., Figure 10.1) which has indisputable mechanistic significance. Such breaks demonstrate a shift in mechanism and the onset of participation. Unfortunately, this tool is a rather coarse one, capable of detecting participation only in systems where anchimeric assistance is very large in magnitude (e.g., in **2** and possibly in **14**, although not enough data points are available). No definite breaks in Hammett–Brown plots are observed with $3 \equiv 8$, **27**, **29** and even with **10**, **16**, and **18**, where π-participation in the secondary derivatives definitely is a factor. The problem is that all the aryl groups, even $3,5\text{-}(CF_3)_2C_6H_3$, are π donors and stabilize the classical carbocation preferentially. Such aryl-substituted tertiary cations are classical, but their secondary counterparts may well be bridged.

What magnitude of anchimeric assistance is needed for the Gassman–Fentiman criterion to be operative? A break in the Hammett–Brown plot (Fig. 10.1) is found for **2**, but the secondary derivative exhibits 10^{11} in π-assistance. No break in the plot for **18** is found, despite 10^5–10^6 of assistance in the secondary system.

---

[42]Tanida and I have discussed possible interpretations.[41] *Added in proof*: W. L. Jorgensen (private communication) has just developed a predictive approach based on perturbation theory and MINDO/3 calculations.

The "tool of increasing electron demand" is too insensitive to detect anchimeric assistance of even moderately large magnitude, and does not seem capable of contributing very much information about the norbornyl problem.[43]

It would indeed "be desirable to vary the electron demand at position 2 while maintaining the steric environment essentially constant" by investigating substituted *secondary* 2-norbornyl systems, **38** and **39**, rather than **33** and **35**, which do not appear to me to be ideal for this purpose.

	38	**39**	**40**	
exo:endo (25°C):	3.9	3.2	1240	

The electron-donating exocyclic methylene of **38** and the cyclopropane ring of **39** bisect the 2-position as does the *gem*-dimethyl group of **40**. The *exo:endo* ratio of **40** (Table 10.1) is comparable to that of the parent norbornyl system, whereas **38** and **39** give very low values.

Figures 10.2 and 10.4 are misleading. Were data for **38**, **39** and many other systems to be plotted, entirely different patterns of behavior would be revealed. I see no objective reason for omitting points for **38** and **39**, but including those for **31**, **33**, **34**, and **35**.

The 3-tosylate substituents for **33** and **35** do not bisect the 2-position, and do not maintain "the steric environment essentially constant." Models like **31** and **34** provide an uncertain calibration. Anchimerically assisted solvolysis of **34** distributes the charge from C2 to C5, but solvolysis of **35** would produce quite a different charge delocalization from C2 and C1. How does this affect the results? The dipolar effect of the 3-tosylate substituent adjacent to the reaction site also introduces serious complications, as has already been pointed out by Kleinfelter and Miller.[44] Such dipolar effects can be reduced by placing substituents at more remote positions, and this strikes me as being a sounder approach (see data in Chapter 11).

Unfortunately, one can have little confidence in the reliability of conclusions drawn from Peters' method.[40] Closer analysis reveals many

---

[43]While the solvolytic *exo:endo* ratios of 2-aryl-2-norbornyl derivatives do not reveal a significant trend when the aryl groups become less electron demanding, such trends are found in the ^{1}H, ^{19}F, and ^{13}C chemical shifts of the corresponding stable tertiary cations. Such nmr probes may be more sensitive than solvolysis for the detection of "the onset of nonclassical stabilizations". D. G. Farnum and A. D. Wolf, *J. Amer. Chem. Soc.*, **96**, 5166 (1974); G. A. Olah, G. K. S. Prakash, and G. Liang, *J. Amer. Chem. Soc.*, submitted; D. G. Farnum, R. E. Botto, W. T. Chambers, and B. Lam, to be published.

[44]D. C. Kleinfelter and J. M. Miller, Jr., *J. Org. Chem.*, **38**, 4142 (1973).

problems. While it is remarkable that solvolysis rate constants of secondary derivatives can be estimated at all from tertiary aryl data, the accuracy is not high. Deviations of factors 7 or more are found in half the systems examined; comparison involving two compounds could well be in error by a factor of 50 or more. Using available data (Table 10.1), an error of over 40 for cyclohexyl is found; the plot resembles Figure 10.8. Are we to conclude that cyclohexyl is anchimerically assisted? Acetolysis data on secondary tosylates was employed by Peters; solvent assistance plays an important role (10^2 or more) in HOAc for some of the systems treated, but not for others.[45] This introduces further uncertainty.

A similar approach has been used to compare the more closely related tertiary systems, methyl- with aryl-substituted. Even though the extrapolations involved are not as large as with Peters' cases, numerous compounds are not correlated satisfactorily.[46]

Finally, let us test objectively the predictive power of Peters' method with the newly investigated Coates system, **41**.[47] This system affords the apparent advantage that all measurements, R = H and aryl, were made directly in the same solvents with no need to correct for different leaving groups.

41

In 65% acetone at 25°C, Coates and Fretz find $(R = \phi)/(R = H) = 1600$. In 80% acetone, Brown and Ravindranathan obtain a similar ratio of 1800, and $\rho+ = -2.05$ for R = aryl. According to Peters' treatment, $(R = \phi)/(R = H)$ should be $10^{2.05 \times 2.53}$ or 154,000. The discrepancy with experiment approaches 10^2 and is unacceptably large. Regrettably, I must conclude that Peters' method is not to be trusted in its present stage of development.[48]

---

[45]F. L. Schadt, T. W. Bentley, and P. v. R. Schleyer, *J. Amer. Chem. Soc.*, **98**, 7667 (1976).

[46]J. M. Harris and S. P. McManus, private communication. E. N. Peters, private communication.

[47]In an important recent development, the "tool of increasing electron demand" has confirmed the presence of σ-bridging in the Coates' cation (**44**, Chapter 14). H. C. Brown and M. Ravindranathan, *J. Amer. Chem. Soc.*, **99**, 299 (1977); R. M. Coates and E. R. Fretz, *ibid.*, **99**, 297 (1977).

[48]Professor Brown has suggested to me that the problem may lie in the OTs-OPNB conversion factor used in all other systems. Moreover, the constancy of the conversion factors has yet to be established.

[*Note added in proof*] Data recently collected by Brown's group on the solvolysis of aryldialkylcarbinyl (RR'ArCOPNB) and 1-aryl-1-cycloalkyl *p*-nitrobenzoates[49] permit further evaluation of the accuracy of Peters' method. The error between the value calculated and experiment is a factor of 32 for the isopropyl system. Similar error factors are 68 for 3-pentyl, 1800 for 3-methyl-2-butyl, 1070 for cyclobutyl, 0.09 for cyclopentyl, and 49 for cyclohexyl. While the cycloheptyl, cyclooctyl, and 2-bicyclo[2.1.1]hexyl[50] systems do correlate well (error factors 0.7, 0.6, and 0.4, respectively), this agreement can only be regarded as fortuitous in light of the other results.

[*Reply* (HCB)] The above note is an exquisite example of the art of "sciencemanship." The tertiary aryl derivatives utilized in applying the tool of increasing electron demand must solvolyze by an essentially k_c process, without significant solvent participation. In extrapolating the data to related secondary systems, it would appear desirable to use systems where even the secondary derivatives must involve little or no participation as well as structures reasonably closely related to the bicyclic used to calculate the $\gamma+$ value. These precautions were followed in the correlations exhibited in Figures 10.6–10.9. However, Professor Schleyer applies the extrapolation to systems such as isopropyl, where solvent participation is large (Figure 11.3), and where the structural relationship to the bicyclics is small. He then uses modest discrepancies between the observed and calculated rate constants, 32 for isopropyl, to cast doubt on extrapolations for bicyclic systems where the agreements are extraordinarily good (Figures 10.6, 10.7, and 10.9).

[*Reply* (PvRS)] "Sciencemanship" indeed! For an exquisite example, see Figure 15.1.

[49]H. C. Brown, M. Ravindranathan, E. N. Peters, C. G. Rao, and M. M. Rho, *J. Amer. Chem. Soc.*, in print.
[50]H. C. Brown, M. Ravindranathan, and C. G. Rao, *J. Amer. Chem. Soc.*, **99**, 2359 (1977).

11

Substituent and Structural Effects in 2-Norbornyl

11.1. Introduction

The data for the stabilized 2-*p*-anisyl-2-norbornyl and related tertiary derivatives (Chapter 7) reveal high *exo*:*endo* rate and product ratios comparable to those revealed by the parent secondary system, 2-norbornyl itself. This is made evident by a comparison of the Goering–Schewene diagram for the solvolysis of 1,2-dimethyl-2-norbornyl *p*-nitrobenzoate (Figure 11.1) with that for 2-norbornyl acetate (Figure 11.2).

It is now accepted that such stabilized tertiary 2-norbornyl cations are classical.[1] However, it is argued that the secondary 2-norbornyl cation is nonclassical, with a σ-bridge.[1] Is it reasonable to ascribe such remarkably similar behavior (Figures 11.1 and 11.2) to the operation of totally different physical factors? Sargent evidently found the effects of substituents and structural modifications in the secondary 2-norbornyl system highly persuasive. It is time we consider these results.

11.2. Secondary vs. Tertiary 2-Norbornyl Cations

Sargent has taken the position that it is not unreasonable to ascribe the high *exo*:*endo* rate ratios in tertiary 2-norbornyl derivatives to steric hindrance to ionization and the similarly high *exo*:*endo* rate ratios in secondary 2-norbornyl derivatives to σ-bridging.[1] He justifies this position on the basis of the difference in the steric requirements of a 2-alkyl or 2-aryl

---

[1]G. D. Sargent, *Carbonium Ions*, G. A. Olah and P. v. R. Schleyer, Eds., Vol. III, Wiley–Interscience, New York, 1972, Chapter 24.

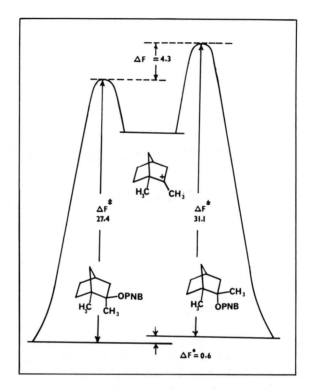

Figure 11.1. Goering–Schewene diagram for the solvolysis of the 1,2-dimethyl-*exo*- and -*endo*-norbornyl *p*-nitrobenzoates in 80% acetone.

substituent with that of a 2-hydrogen atom. He considers that in the tertiary derivative the 2-alkyl or 2-aryl substituent will assist in trapping the anion in the *endo* cavity. However, he believes that the 2-hydrogen atom in the secondary derivative is too small to contribute to such trapping.

An examination of molecular models suggests that the 2-hydrogen atom in the secondary norbornyl group can contribute to the trapping of the anion in the *endo* cavity (Figure 8.3). Further weakening the argument is the observation that the experimental *exo*: *endo* rate ratios do not vary greatly for secondary 2-norbornyl,[2] tertiary 2-methyl-2-norbornyl,[3] tertiary 2-*tert*-butyl-2-norbornyl,[4] and tertiary 2-*p*-anisyl-2-norbornyl[5] (1).

[2]S. Winstein and D. Trifan, *J. Amer. Chem. Soc.*, **74**, 1147, 1154 (1952).
[3]S. Ikegami, D. L. Vander Jagt, and H. C. Brown, *J. Amer. Chem. Soc.*, **90**, 7124 (1968).
[4]E. N. Peters and H. C. Brown, *J. Amer. Chem. Soc.*, **96**, 265 (1974).
[5]H. C. Brown and K. Takeuchi, *J. Amer. Chem. Soc.*, **90**, 2691 (1968).

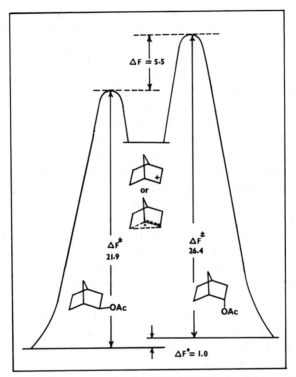

Figure 11.2. Goering–Schewene diagram for the acid-catalyzed acetolyses of *exo*- and *endo*-norbornyl acetates.

OBs / H	OPNB / CH₃	OPNB / H₃C—C—CH₃ / CH₃	OPNB / OCH₃ (1)

exo:*endo*:   350 k_t 885   470   284
          1600 k_α

A more serious issue would appear to be the question as to whether a true carbonium ion intermediate is involved in the solvolysis of such secondary derivatives. "Of fundamental importance in the S_N1 mechanism is the postulate of a carbonium ion intermediate."[6] Such carbonium ion

[6] A. Streitwieser, Jr., *Solvolytic Displacement Reactions*, McGraw-Hill, New York, 1962.

intermediates have been demonstrated in a number of solvolytic processes involving tertiary derivatives,[6] but the situation is far less clear for secondary derivatives. Thus Hughes and Ingold considered the solvolysis in aqueous solvents of isopropyl chloride and similar secondary halides to represent a mechanistic transition point between S_N2 and S_N1 behavior.[7,8] However, it was later proposed that the solvolysis of secondary derivatives could be made essentially "limiting"[9] (k_c) by shifting to better leaving groups, e.g., arenesulfonates, and by utilizing less nucleophilic solvents, such as acetic acid and formic acid. Consequently, the acetolysis of secondary arenesulfonates has long been emphasized in studies of structural effects in solvolysis.[6]

More recently it has been concluded that such solvolyses often involve a large contribution from solvent[10,11] (k_s). It is proposed that the solvolysis of 2-adamantyl tosylate involves a true k_c process ($m^9 = 0.91$, the highest value found for a secondary aliphatic tosylate) with the adamantyl structure protecting the backside of the developing cation from solvent participation.[10] Consequently, a comparison of the effect of changes in solvent on the rate of solvolysis of a particular substrate with the effect of 2-adamantyl tosylate provides a measure of the solvent contribution to the solvolysis of the former.

It has also been proposed that solvolysis of secondary derivatives in hexafluoroisopropyl alcohol (HFIP) and trifluoroacetic acid (TFA) is essentially limiting[12] (k_c). Solvolysis with participation (k_Δ) should also be insensitive to solvent participation.

The behavior of a representative secondary system, 2-propyl tosylate ($m = 0.44$), is exhibited in Figure 11.3. Clearly, we are not dealing here with a true k_c process. Such solvolytic processes are better termed "carbonoid," rather than "carbonium ion."

The question is whether it is reasonable to treat the solvolysis of *exo*-norbornyl tosylate ($m = 0.75$) and *endo*-norbornyl tosylate ($m = 0.69$) as proceeding through a true carbonium ion intermediate. Indeed, we find that the available data for both *exo*- and *endo*-norbornyl tosylate are well correlated in this way (Figures 11.4 and 11.5). Consequently, it appears that

[7]L. C. Bateman, M. G. Church, E. D. Hughes, C. K. Ingold, and N. A. Taher, *J. Chem. Soc.*, 979 (1940).

[8]C. K. Ingold, *Structure and Mechanism in Organic Chemistry*, 2nd ed., Cornell University Press, Ithaca, N.Y., 1969, Section 27.

[9]S. Winstein, E. Grunwald, and H. W. Jones, *J. Amer. Chem. Soc.*, **73**, 2700 (1951).

[10]J. L. Fry, C. J. Lancelot, L. K. M. Lam, J. M. Harris, R. C. Bingham, D. J. Raber, R. E. Hall, and P. v. R. Schleyer, *J. Amer. Chem. Soc.*, **92**, 2538 (1970); J. L. Fry, J. M. Harris, R. C. Bingham, and P. v. R. Schleyer, *ibid.*, **92**, 2540 (1970); P. v. R. Schleyer, J. L. Fry, L. K. M. Lam, and C. J. Lancelot, *ibid.*, **92**, 2542 (1970).

[11]D. J. Raber and J. M. Harris, *J. Chem. Ed.*, **49**, 60 (1972).

[12]F. L. Schadt, P. v. R. Schleyer, and T. W. Bentley, *Tetrahedron Lett.*, 2335 (1974).

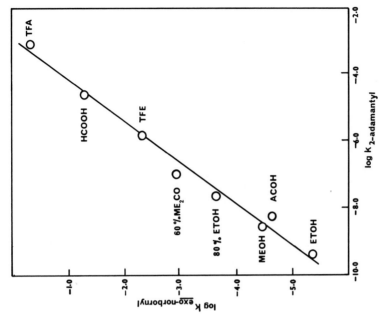

Figure 11.4. Existence of a free-energy correlation for the solvolysis of *exo*-norbornyl tosylate in representative solvents.

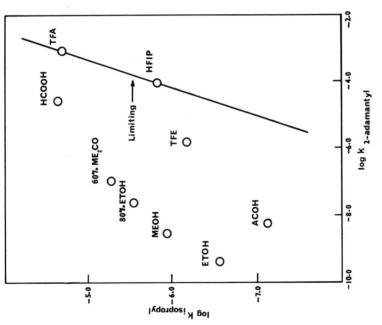

Figure 11.3. Absence of a free-energy correlation for the solvolysis of isopropyl tosylate in representative solvents.

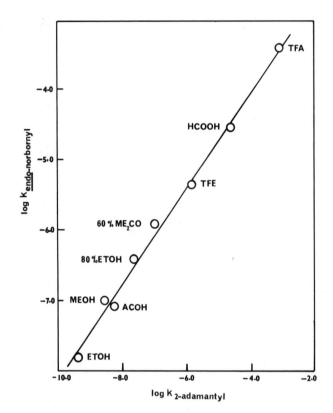

Figure 11.5. Existence of a free-energy correlation for the solvolysis of *endo*-norbornyl tosylate in representative solvents.

both *exo*- and *endo*-norbornyl tosylates undergo solvolysis to produce a carbonium ion (or ion pair) intermediate encumbered very weakly, if at all, by solvent. In view of the excellence of this correlation, it is of interest that at the present time the solvolysis of these derivatives is interpreted as a k_Δ process for *exo*-,[13] a k_s process for *endo*-,[13] and a k_c process for 2-adamantyl.[10] Yet they correlate very well (Figures 11.4 and 11.5).

The data are summarized in Table 11.1.

The existence of the Goering–Schewene diagram for the secondary 2-norbornyl system (Figure 11.2) and the tertiary system (Figure 11.1) likewise supports the conclusion that the solvolysis of both systems involve the formation of true carbonium ion intermediates. Moreover, the observa-

[13]J. E. Nordlander, R. R. Gruetzmacher, W. J. Kelly, and S. P. Jindal, *J. Amer. Chem. Soc.*, **96**, 181 (1974).

TABLE 11.1. *Rates of Solvolysis of Tosylates in Representative Solvents*

Solvent	$10^6 k_1$, sec^{-1} at 25°C			
	2-Propyl	*endo*-Norbornyl	*exo*-Norbornyl	2-Adamantyl[o]
Ethanol	0.286[a]	0.0156[j]	4.39[n]	0.00043
Acetic acid	0.0774[b]	0.0828[k]	23.3[k]	0.0059
Methanol	1.18[c]	0.102[c]	34.8[c]	0.0029
80% Ethanol	2.94[a,b]	0.397[l]	231.0[l]	0.024
60% Acetone	5.56[c,d]	1.13[c,d]	1,190[c,d]	0.111
97% Trifluoro-ethanol[e]	0.692[f,g]	4.6[g]	4,750[g]	1.64
Formic acid	23.8[h]	30.1[m]	51,250[m]	27.5
Trifluoroacetic acid	21.4[i]	417[i]	468,000[i]	900

[a] R. E. Robertson, *Can. J. Chem.*, **31**, 589 (1953).   [b] P. v. R. Schleyer et al., *J. Amer. Chem. Soc.*, **92**, 2542 (1970).   [c] Unpublished research with F. J. Chloupek.   [d] Calculated from the rate of mesylate using the ratio $k_{OTs} = k_{OMs} \times 2$.   [e] 3% water by weight.   [f] Calculated from the rate of brosylate using the ratio $k_{OBs} = k_{OTs} \times 3$.   [g] J. M. Harris, personal communication.   [h] P. E. Peterson et al., *J. Amer. Chem. Soc.*, **87**, 5169 (1965).   [i] J. E. Nordlander et al., *J. Amer. Chem. Soc.*, **96**, 181 (1974).   [j] Calculated using *exo* : *endo* = 280.   [k] P. v. R. Schleyer et al., *J. Amer. Chem. Soc.*, **87**, 375 (1965).   [l] D. Lenoir, *Chem. Ber.*, **108**, 2055 (1975).   [m] Unpublished research with I. Rothberg.   [n] M.-H. Rei, Ph.D. Thesis, Purdue University, 1967.   [o] F. L. Schadt, T. W. Bentley, and P. v. R. Schleyer, *J. Amer. Chem. Soc.*, **98**, 7667 (1976).

tion that the *exo* : *endo* product ratios realized are correlated with the differences in the free energies of the two transition states (Figures 11.1, 11.2, 7.1, 7.2, and 8.6–8.9) supports the conclusion that the solvolysis of both the *exo*- and *endo*-derivative proceeds through a common carbonium ion intermediate which distributes itself to product in accordance with the relative energies of the two transition states.

It should be recalled, however, that the acetolysis of optically active *endo*-norbornyl brosylate forms 7% active *exo*-norbornyl acetate, whereas the *exo* isomer yields only inactive *exo* acetate.[2] Consequently, the two products cannot be coming from an identical common intermediate. It is convenient to ignore this deviation of 7%, but it does imply that we are dealing with a borderline situation which might easily be shifted by substituents or structure modifications to a process in which the products from *exo* and *endo* are clearly different, no longer attributable to a common intermediate. Such solvolyses will no longer provide a simple Goering–Schewene diagram.

It is unfortunate that many of the arguments used to support σ-bridged structures[1] have ignored this point and have failed even to consider whether the solvolysis of *exo*- and *endo*-derivatives proceed through a common carbonium ion intermediate.

11.3. Substituents as a Probe for Charge Delocalization

The use of substituents as a probe for charge delocalization has been a favorite, highly reliable tool of organic chemists.[8] Let us consider two typical applications.

The fast rate of solvolysis of *tert*-cumyl chloride as compared to *tert*-butyl chloride (2) is attributed to stabilization of the incipient cation in

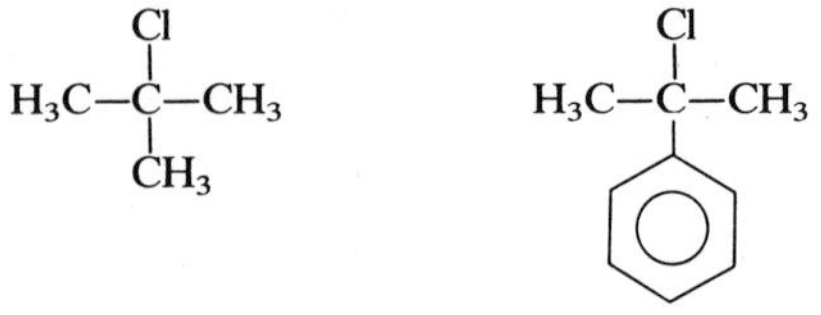

$$
\begin{array}{ccc}
\underset{\underset{\text{CH}_3}{|}}{\overset{\overset{\text{Cl}}{|}}{\text{H}_3\text{C}-\text{C}-\text{CH}_3}} & \overset{\overset{\text{Cl}}{|}}{\text{H}_3\text{C}-\text{C}-\text{CH}_3} & \qquad (2)
\end{array}
$$

RR (25°C):   1.00   6000

the transition state through charge delocalization into the aromatic ring (3). (For convenience, the fully formed cation is represented.)

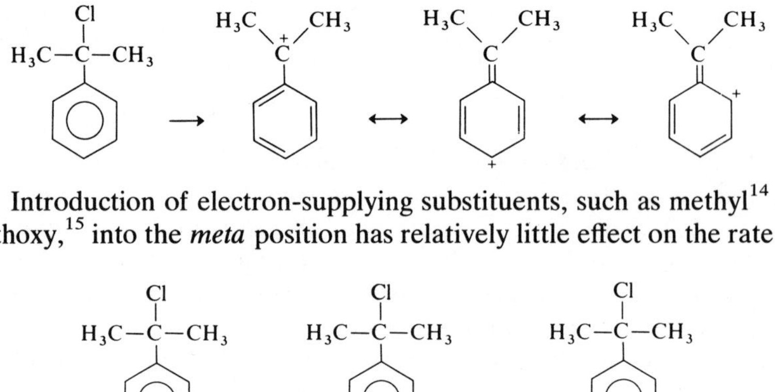

(3)

Introduction of electron-supplying substituents, such as methyl[14] and methoxy,[15] into the *meta* position has relatively little effect on the rate (4).

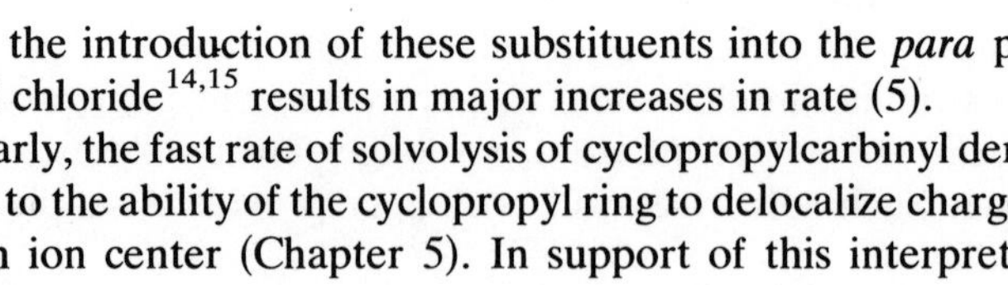

(4)

RR (25°C):   1.00   2.00   0.61

However, the introduction of these substituents into the *para* position of *tert*-cumyl chloride[14,15] results in major increases in rate (5).

Similarly, the fast rate of solvolysis of cyclopropylcarbinyl derivatives is attributed to the ability of the cyclopropyl ring to delocalize charge from the carbonium ion center (Chapter 5). In support of this interpretation, the

[14]H. C. Brown, J. D. Brady, M. Grayson, and W. H. Bonner, *J. Amer. Chem. Soc.*, **79**, 1897 (1957).
[15]Y. Okamoto and H. C. Brown, *J. Amer. Chem. Soc.*, **79**, 1909 (1957).

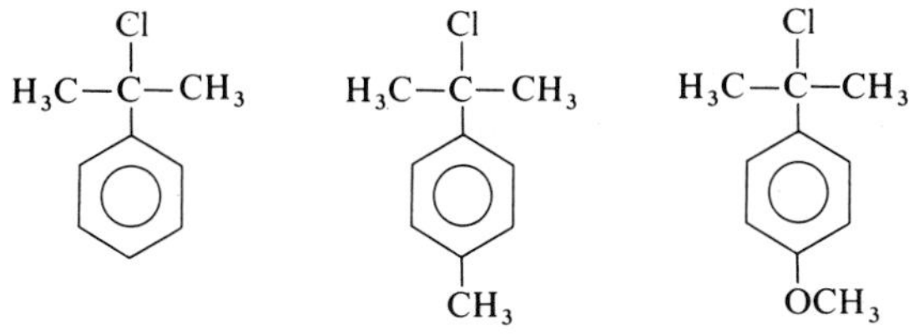

(5)

RR (25°C):	1.00	26	3360

introduction of the methyl or the ethoxy substituent results in large rate increases (6).[16]

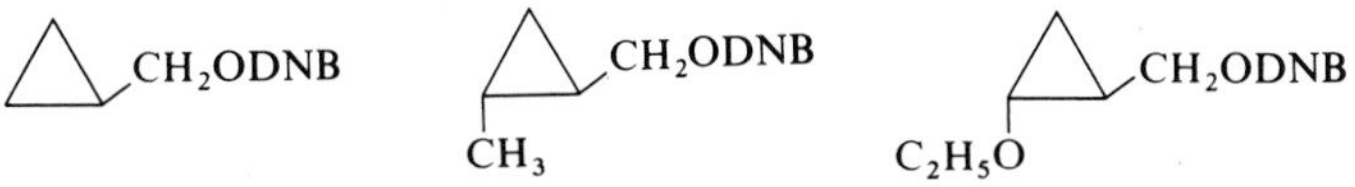

(6)

RR (25°C):	1.00	11.0	940

Accordingly, it appears appropriate to apply the same tool to 2-norbornyl to test for the often postulated charge delocalization from the 2-position.

11.4. The Search for Charge Delocalization at C6

The nonclassical 2-norbornyl cation (**1**) was originally formulated as a resonance hybrid (7) of three canonical structures (**1′, 1″, 1‴**)^2:

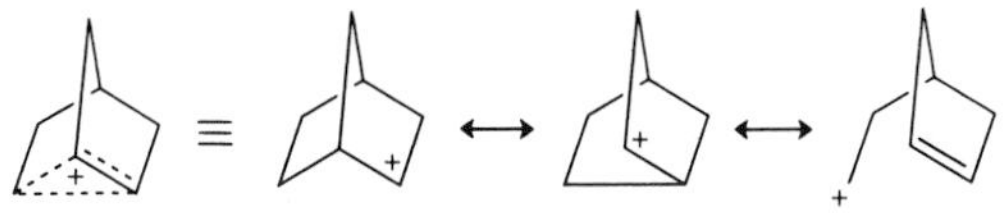

(7)

On this basis charge should be delocalized from the 2- to the 1- and 6-positions. If so, the introduction of electron-supplying substituents, such as methyl and methoxy, into the 6-position would be expected to result in stabilization of the 2-norbornyl cation and in rate increases, such as were observed in the *tert*-cumyl (5) and cyclopropylcarbinyl (6) systems.

McGreer examined the addition of acetic acid to 6,6-dimethylnorbornene (**2**)[17] (8). He had anticipated that protonation would

[16]P. v. R. Schleyer and G. W. Van Dine, *J. Amer. Chem. Soc.*, **88**, 2321 (1966).
[17]D. E. McGreer, *Can. J. Chem.*, **40**, 1554 (1962).

occur preferentially at C3 to give the more stable nonclassical ion **3** leading to the preferred formation of the 2-acetate **4**. In fact, he observed a considerable preference for protonation at C2 (**5**) to give **6** (**6:4** = 2 : 1).

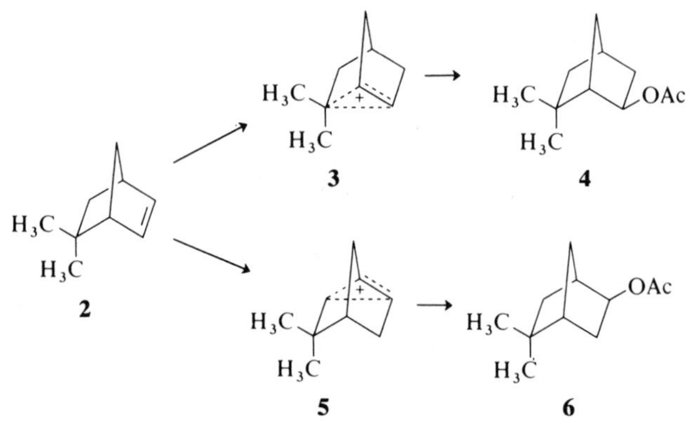

(8)

One can, of course, argue that protonation is not solvolysis, so that the true nonclassical ion is formed only in solvolysis.[1] However, the introduction of one (**8**)[18] or two[19] methyl groups (**9**) or a methoxy group (**10**)[20] into the 6-position of 2-norbornyl tosylate (**7**) fails to reveal any increase in the rate of acetolysis (**9**).

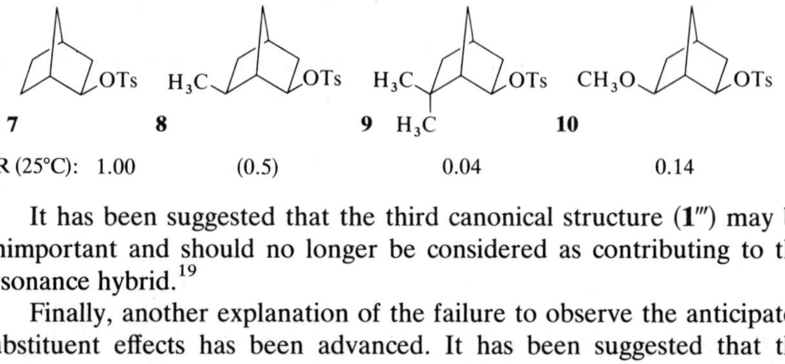

(9)

RR (25°C):   1.00   (0.5)   0.04   0.14

It has been suggested that the third canonical structure (**1‴**) may be unimportant and should no longer be considered as contributing to the resonance hybrid.[19]

Finally, another explanation of the failure to observe the anticipated substituent effects has been advanced. It has been suggested that the interaction of substituents with electron deficiencies in nonclassical cations may be far smaller than those observed in the *tert*-cumyl (5) and cyclopropylcarbinyl (6) cations.[21]

[18]Private communication from P. v. R. Schleyer.

[19]P. v. R. Schleyer, M. M. Donaldson, and W. E. Watts, *J. Amer. Chem. Soc.*, **87**, 375 (1965).

[20]P. v. R. Schleyer, P. T. Stang, and D. J. Raber, *J. Amer. Chem. Soc.*, **92**, 4725 (1970).

[21]C. F. Wilcox, Jr., L. M. Loew, R. G. Jesaitis, S. Belin, and J. N. C. Hsu, *J. Amer. Chem. Soc.*, **96**, 4061 (1974).

However, it is a basic tenet in science that it must be possible to test independently for the effects proposed. Thus far all direct tests for charge delocalization from the 2- to the 6-position have failed.

11.5. The Search for Charge Delocalization at C2

Perhaps a more direct approach would be to test for charge delocalization from C2. If the 2-norbornyl cation is indeed a resonance stabilized species, the effect of a methyl or phenyl group at C2 would be considerably smaller than its effect upon a representative system not so stabilized.

For example, the introduction of a methyl group in isopropyl chloride, forming *tert*-butyl chloride, results in a calculated rate enhancement of ethanolysis by a factor of 55,000 (10).[22]

$$\begin{array}{ccc} \overset{\text{H}}{\underset{\text{Cl}}{H_3C-\overset{|}{\underset{|}{C}}-CH_3}} & \overset{\text{CH}_3}{\underset{\text{Cl}}{H_3C-\overset{|}{\underset{|}{C}}-CH_3}} & (10) \end{array}$$

RR (25°C):   1.00   55,000

(In this estimate an attempt was made to minimize the effect of solvent participation by using a "calculated" rate for the secondary chloride derived from the rate of acetolysis of the tosylate.)

The introduction of a methyl group into the central carbon atom of benzhydryl chloride results in a calculated rate enhancement by a much smaller factor, 346 (11).

$$\begin{array}{ccc} \overset{\text{H}}{\underset{\text{Cl}}{\bigcirc-\overset{|}{\underset{|}{C}}-\bigcirc}} & \overset{\text{CH}_3}{\underset{\text{Cl}}{\bigcirc-\overset{|}{\underset{|}{C}}-\bigcirc}} & (11) \end{array}$$

RR (25°C):   1.00   346

This smaller factor is attributed to a "leveling effect." The stabilized benzhydryl cation makes a much smaller demand upon the new methyl substituent for additional stabilization than does the relatively hungry, less stabilized isopropyl cation. Consequently, if the 2-norbornyl cation is really a resonance stabilized species, the effect of substituents in the 2-position should be greatly decreased over their effect in a typical aliphatic or alicyclic system. However, very similar effects are observed (Table 11.2).

An alternative approach which does not require utilization of the questionable secondary derivatives is the calculation of k_{Ph}/k_{Me} for tertiary

[22]H. C. Brown and M.-H. Rei, *J. Amer. Chem. Soc.*, **86**, 5008 (1964).

derivatives.[22] Here also we observe a diminishing effect of the phenyl substituent relative to the methyl substituents for the more stabilized systems (12).

$$H_3C-\underset{\underset{Cl}{|}}{\overset{\overset{R}{|}}{C}}-CH_3 \qquad \underset{\underset{Cl}{|}}{\overset{\overset{R}{|}}{C}}-CH_3 \qquad \underset{\underset{Cl}{|}}{\overset{\overset{R}{|}}{C}} \tag{12}$$

k_{Ph}/k_{Me} (25°C):   4580   50   29

Again we observe that Ph/Me is 5260 in 2-norbornyl (Table 11.2).

We have previously pointed out that the solvolysis of both *exo*- and *endo*-norbornyl tosylates correlate well with 2-adamantyl tosylate (Figures

TABLE 11.2. *Effect of Substituents on the Rate of Solvolysis*

Acyclic derivative	Relative rate	2-Norbornyl derivative	Relative rate
$\underset{\underset{CH_3,}{|}}{\overset{\overset{CH_3}{|}}{HC}}-X$	1.00		1.00
$\underset{\underset{CH_3}{|}}{\overset{\overset{CH_3}{|}}{H_3C-C}}-X$	55,000		55,000
$\underset{\underset{CH_3}{|}}{\overset{\overset{CH_3}{|}}{C}}-X$	(×4,600) 2.6×10^8		(×5,300) 2.9×10^8
$H_3CO-\underset{\underset{CH_3}{|}}{\overset{\overset{CH_3}{|}}{C}}-X$	(×3,400) 8.5×10^{11}		(×1,700) 5.0×10^{11}

11.4 and 11.5). Consequently, solvent participation cannot be a significant factor. Let us compare k_{Me}/k_H for both *exo*- and *endo*-norbornyl. If the *exo* isomer is strongly resonance stabilized, then k_{Me}/k_H must be much smaller than the value realized in the *endo* isomer.

We have data for the secondary tosylates in acetic acid[19] and for the tertiary *p*-nitrobenzoates in 80% acetone.[23] Instead of attempting to calculate a conversion factor from the tosylate in acetic acid to the *p*-nitrobenzoate in 80% acetone, let us adopt the value *c* for this conversion factor.

The data for the *exo* derivatives (13) lead to a value of $(k_{Me}/k_H)_{exo} = 4.29 \times 10^{-4} c^{-1}$.

$$\tag{13}$$

| k_t (25°C): | 2.33×10^{-5} | $2.33 \times 10^{-5} c$ | 1.00×10^{-8} |

Similarly, the data for the *endo* derivatives (14) lead to a value of $(k_{Me}/k_H)_{endo} = 1.37 \times 10^{-4} c^{-1}$.

$$\tag{14}$$

| k_t (25°C): | 8.28×10^{-8} | $8.28 \times 10^{-8} c$ | 1.13×10^{-11} |

The results clearly reveal that k_{Me}/k_H is comparable for the two isomers: $(k_{Me}/k_H)_{exo}/(k_{Me}/k_H)_{endo} = 3.1$. That is, k_{Me}/k_H for *exo*-norbornyl is not much smaller than the value for the *endo* system, but is slightly larger. If we use k_α $(4.6k_t)$ for the *exo* tosylate, the value of $(k_{Me}/k_H)_{exo}/(k_{Me}/k_H)_{endo} = 0.67$. Clearly, the application of this test reveals no evidence for significant charge delocalization from the 2-position that is different in the *exo* isomer than in the *endo*. (It should be recognized that this is merely another way of restating the observation that the *exo*:*endo* rate ratios are very similar in the secondary and in the tertiary derivatives.) For an examination of k_{Me}/k_H values, using the extreme limiting condition of solvolysis of the secondary tosylates in trifluoroacetic acid, see Section 14.4.

[23]S. Ikegami, D. L. Vander Jagt, and H. C. Brown, *J. Amer. Chem. Soc.*, **90**, 7124 (1968).

11.6. The Search for Charge Delocalization at C1

There is a difficulty in introducing substituents into the 1-position as a test for charge delocalization in the parent system. Such 2-norbornyl derivatives containing alkyl or aryl groups in the 1-position solvolyze with rearrangement to tertiary 2-norbornyl cations:

$$\text{(15)}$$

Consequently, such data can reveal that partial rearrangement to the more stable cation is occurring in the transition state, but they cannot provide an answer to the question of whether charge is being delocalized to the 1-position in a system, such as 2-norbornyl itself, which is not undergoing solvolysis with rearrangement to a more stable structure (degenerate rearrangement).

Nevertheless, an examination of the data reveal major differences in the effect of substituents in the 1- and 2-positions (Table 11.3). The results suggest that the transition state is only modestly along the reaction coordinate toward the tertiary cation.

Since the 1- and 2-substituted derivatives doubtless undergo solvolysis without internal return, the observed rates should be compared to 2-norbornyl corrected for internal return. This correction, if valid, reduces the

TABLE 11.3. Effect of 1- and 2-Substituents on Rates of Solvolysis

R	![structure 1]	![structure 2]	![structure 3]
Methyl	1.00	50^a $(10)^c$	$55,000^b$ $(11,000)^c$
Ph	1.00	3.9^a $(0.8)^c$	$290,000,000^b$ $(60,000,000)^c$
p-An	1.00	7.8^a $(2.0)^c$	$500,000,000,000^b$ $(100,000,000,000)^c$

[a] D. C. Kleinfelter, Ph.D. Thesis, Princeton University, 1960.
[b] Reference 22.
[c] Corrected for internal return.

effect of the substituents by a factor of approximately 5, as shown by the values in parentheses in Table 11.3.

It has recently been reported that the introduction of a $-CH_2Sn(CH_3)_3$ substituent in the 1-position (**11**) accelerates the rate of solvolysis of *exo*-2-norbornyl *p*-nitrobenzoate by a factor of 6×10^5.[24] The reaction proceeds with both rearrangement and elimination (**12**) (16).

$$
\text{11} \quad \longrightarrow \quad \text{12} \quad + \ (CH_3)_3SnX \tag{16}
$$

This is a major enhancement in the rate. Regrettably, it was not established that the reaction proceeds through a true carbonium ion. The pertinence of this result to the question of σ-bridging in a system which is not rearranging or rearranging degenerately is not clear.

The introduction of a spiro cyclopropane moiety at the 7-position (**13**) should stabilize developing positive charge at the 1-position (17). The

$$
\text{13} \tag{17}
$$

RR (25°C):   3.4   1.00

compound also solvolyzes with rearrangement (18), but the observed rate is only faster than that of the parent compound by a factor of 3.4.[25] Consequently, even here it is not possible to say how much, if any, of the small rate

$$
\xrightarrow{\text{HOAc}} \tag{18}
$$

enhancement is the result of charge delocalization to the 1-position which is not associated with the rearrangement.

Ideally one should test for charge delocalization with substituents by introducing them in a system which does not rearrange or undergoes only degenerate rearrangement. Thus the introduction of a 1-methyl group into 2-methylnorbornyl exhibits identical effects both in the *endo* and *exo*

[24]G. D. Hartman and T. G. Traylor, *J. Amer. Chem. Soc.*, **97**, 6147 (1975).
[25]C. F. Wilcox, Jr., and R. G. Jesaitis, *Tetrahedron Lett.*, 2567 (1967).

isomers (Section 9.5). Clearly charge is not being delocalized to the 1-position in the solvolysis of 2-methyl-*exo*-norbornyl *p*-nitrobenzoate.

Phenyl, *p*-anisyl, and cyclohexyl substituents have been introduced into the 1-, 3-, and 7-position of 2-norbornyl tosylate. The effects on the rate are interesting, but do not require σ-bridging to account for the results.[26]

Data on the effect of such substituents on the rates of solvolysis of 2-norbornyl derivatives are summarized in Table 11.4.

[26]D. C. Kleinfelter et al., *J. Org. Chem.*, **38**, 4127, 4134, 4142 (1973).

TABLE 11.4. Effect of Activating Substituents on Rates of Solvolysis of 2-Norbornyl Derivatives

Compound	X	$10^6 k_1$, sec^{-1} at 25°C	$\Delta H^{\ddagger}$, kcal mol^{-1}	$\Delta S^{\ddagger}$, eu	RR at 25°C	$\dfrac{exo}{endo}$	Reference
(norbornyl structure, positions 1–7, X at 2)	X-OBs	88.2	25.1	7.3	1.00	350	a
	N-OBs	0.252	26.0	−1.5	1.00		
	X-OTs	23.3	21.6	−7.2	1.00	280	19
	N-OTs	0.0828	25.8	−4.4	1.00		
1-Me	X-OTs	1250			51.2	14,100	b
	N-OTs	0.0932	25.8	−4.1	1.13		
1-Et	X-OTs	1900			77.8	22,900	b
	N-OTs	0.0923	27.2	0.4	1.12		
1-CyHex	X-OTs	3510			150	4,800	26
	N-OTs	0.732	24.8	−3.5	8.84		
1-Ph	X-OTs	95.5	22.8	−0.3	3.91	1,690	26
	N-OTs	0.0566	25.1	−7.7	0.683		
1-*p*-An	X-OTs	188	22.1	−1.6	7.7	2,270	26
	N-OTs	0.0623	25.5	−6.1	0.752		
1-OMe	X-OTs	29.8	22.3	−4.5	1.28	6,200	20
	N-OTs	0.00482	26.1	−9.0	0.058		
exo-3-CyHex	X-OTs	25.0	23.8	0.14	1.07	3,380	26
	N-OTs	0.0074	28.3	−0.72	0.087		
endo-3-CyHex	X-OTs	36.4	22.4	−3.6	1.56	21	26
	N-OTs	1.70	26.6	4.1	20.5		
exo-3-Ph	X-OTs	0.182	25.5	−3.8	0.0078	570	26
	N-OTs	0.000319	29.5	−3.0	0.00385		
endo-3-Ph	X-OTs	6.93	23.3	−4.0	0.298	36	26
	N-OTs	0.191	25.3	−4.5	2.31		
syn-7-CyHex	X-OTs	43.7	23.1	−0.91	1.88	710	26
	N-OTs	0.0617	27.1	−0.72	0.745		
anti-7-CyHex	X-OTs	27.3	23.3	−1.6	1.17	244	26
	N-OTs	0.112	26.2	−2.6	1.35		
syn-7-Ph	X-OTs	6.04	23.5	−3.6	0.259	202	26
	N-OTs	0.0299	25.8	−6.5	0.361		

TABLE 11.4 (continued)

Compound	X	$10^6 k_1$, sec^{-1} at 25°C	$\Delta H^{\ddagger}$, kcal mol^{-1}	$\Delta S^{\ddagger}$, eu	RR at 25°C	$\dfrac{exo}{endo}$	Reference
anti-7-Ph	X-OTs	5.68	24.5	−0.2	0.244	242	26
	N-OTs	0.0234	26.6	−4.4	0.283		
exo-5-OMe	X-OTs	0.327	26.5	−0.8	0.014	92	20
	N-OTs	0.00354	28.8	−4.7	0.038		
exo-6-OMe	X-OTs	3.30	22.2	−9.1	0.142	101	20
	N-OTs	0.0326	27.7	0.0	0.394		
anti-7-OMe	X-OTs	1.95	23.5	−5.7	0.084	63	20
	N-OTs	0.0308	27.5	−0.8	0.372		
3,3-Me$_2$	X-OBs	32.9			0.373	1,240	c
	N-OBs	0.0266			0.106		
5,5-Me$_2$	X-OBs	26.1			0.296	145	c
	N-OBs	0.180			0.70		
6,6-Me$_2$	X-OTs	0.92	24.0	−5.6	0.0395	206	19
	N-OTs	0.00447	28.5	−1.1	0.054		
7,7-Me$_2$	X-OBs	770			8.73	4,100	c
	N-OBs	0.188			0.746		
3-Methylene	X-OTs	102			4.38	3.9	d
	N-OTs	26.2		316			
3-Spirocyclo-propyl	X-ODNB	0.00182	27.8	−5.3		1.2	25
	N-ODNB	0.00152	25.5	−13.3			
7-Spirocyclo-propyl	X-OTs	78.8	20.9	−7.3	3.39	1,790	25
	N-OTs	0.044	26.1	−4.4	0.53		

[a] S. Winstein, B. K. Morse, E. Grunwald, H. W. Jones, J. Corse, D. Trifan, and H. Marshall, *J. Amer. Chem. Soc.*, **74**, 1127 (1952).
[b] D. C. Kleinfelter, Ph.D. Thesis, Princeton University, 1960.
[c] A. Colter, E. C. Friedrich, N. J. Holness, and S. Winstein, *J. Amer. Chem. Soc.*, **87**, 378 (1965).
[d] C. F. Wilcox, Jr., and R. G. Jesaitis, *Chem. Commun.*, 1046 (1967).

It is evident that the classic tool of the organic chemist, the introduction of electron-supplying substituents, has failed to detect charge delocalization from the 2-position to the 1- and 6-positions. Only in systems which are undergoing solvolysis with rearrangement do we note a significant effect of such substituents at the 1-position. Clearly we must look elsewhere for support of the σ-bridged structure.

11.7. Deuterium as a Substituent

Deuterium substitution in the 2-norbornyl system has been employed for the study of σ-bridging in the solvolysis of such derivatives.[27] The results

[27] D. E. Sunko and S. Borčić, *Isotope Effects in Chemical Reactions*, C. J. Collins and N. S. Bowman, Eds., Van Nostrand–Reinhold, New York, 1970, Chapter 3.

are highly interesting. Unfortunately, it is proving increasingly difficult to achieve an interpretation of secondary isotope effects.[28] The problem is further complicated by the difficulty in making measurements of such secondary isotope effects which stand the test of time.

For example, it was originally reported that k_H/k_D was only 1.10 for 2-*d-exo*-norbornyl brosylate (**14**), but 1.20 for 2-*d-endo*-norbornyl brosylate (**15**) (19).[29] The low k_H/k_D value in the *exo* isomer was attributed to σ-participation.

(19)

	14 D	**15** OBs
$k_H/k_D{}^{29}$	1.10 (30°C)	1.20 (50°C)
$k_H/k_D{}^{27}$	1.21 (25°C)	1.21 (68°C)

However, it now appears that the low value was a consequence of scrambling of the label between the 2- and 1-positions.[27] A redetermination, following the reaction in the initial 3 to 5% of solvolysis, gave a "normal" value. However, the authors conclude that this "normal" value is also consistent with the σ-bridged structure.[27]

Originally it was reported that deuterium in the 3-position gives a k_H/k_D value of 1.01 in the *exo* isomer (**16**) and 1.26 in the *endo* (**16**). This was attributed to charge delocalization in the *exo* transition state (20).[30] However, this value has been revised to 1.10, but here also the revision "does not affect the conclusions."[27]

(20)

	16	**17**	**18**
k_H/k_D (*exo*):	1.10 (44°C)	1.33 (95°C)	1.09 (40°C)
k_H/k_D (*endo*):	1.26 (65°C)	1.31 (121°C)	1.08 (63°C)

The tertiary 2-methyl derivative (**17**) gives k_H/k_D effects even larger than for the *endo* secondary (**16**). This is attributed to the fact that **17**

---

[28]S. E. Scheppele, *Chem. Rev.*, **72**, 511 (1972).

[29]C. C. Lee and E. W. C. Wong, *J. Amer. Chem. Soc.*, **86**, 2752 (1964); *Can. J. Chem.*, **43**, 2254 (1965).

[30]J. M. Jerkunica, S. Borčić, and D. E. Sunko, *Chem. Comm.*, 1489 (1968).

solvolyzes to a classical ion. However, the tertiary 2-phenyl derivative again gives a low k_H/k_D ratio (**18**).[31] What conclusion can be safely drawn?

The introduction of deuterium in the 6-position likewise results in a different secondary isotope effect in the *exo* and *endo* isomers (**21**).[32,33] Again, this is attributed to σ-bridging in the *exo* isomer. However, it is really puzzling that a 2-*d* substituent fails to exhibit any effect, whereas a 6-*d* does, although σ-bridging between the 2- and 6-positions is proposed.

$$ (21) $$

k_H/k_D (*exo*):	1.09 (25°C), 1.15 (44°C)	1.11 (25°C), 1.10 (44°C)
k_H/k_D (*endo*):	0.98 (70°C), 1.02 (65°C)	0.99 (70°C), 1.00 (65°C)

Deuterium in the 6-position of 1,2-dimethyl-*exo*-2-norbornyl *p*-nitrobenzoate (**21**) exhibits almost no effect (**22**).[34] This would fit in with the absence of σ-participation in a classical tertiary cation. However, a negligible effect is also observed in **22**, which solvolyzes with σ-participation and rearrangement[27] (Section 11.6). This is puzzling.

$$ (22) $$

k_H/k_D	1.02	1.03
	(90% acetone, 80°C)	(AcOH, 25°C)

As pointed out in the following Section, the introduction of $-I$ substituents at C7 of the 2-norbornyl system resists the development of positive charge at the adjacent carbon C1. Consequently, σ-bridging should be greatly reduced compared to that present in the parent system. Werstiuk and his coworkers examined the γ-deuterium isotope effects in 7-halo-*endo*-6-deuterium-*exo*-norbornyl brosylates. Instead of the anticipated decrease, he observed a modest increase in k_H/k_D to 1.11–1.13.[35] The authors

[31]J. P. Schaefer, J. P. Foster, M. J. Dagani, and L. M. Honig, *J. Amer. Chem. Soc.*, **90**, 4497 (1968).

[32]B. L. Murr, A. Nickon, T. D. Swartz, and N. H. Werstiuk, *J. Amer. Chem. Soc.*, **89**, 1730 (1967).

[33]J. M. Jerkunica, S. Borčić, and D. E. Sunko, *J. Amer. Chem. Soc.*, **89**, 1732 (1967).

[34]H. L. Goering and K. Humski, *J. Amer. Chem. Soc.*, **91**, 4594 (1969).

[35]N. H. Werstiuk, G. Timmins, and F. P. Cappelli, *Can. J. Chem.*, **51**, 3473 (1973).

conclude that the γ-isotope effect observed in the solvolysis of *exo*-norbornyl brosylate is probably not derived via participation of the C1–C6 bond. (The possibility that the force constants at C6 may be influenced by the developing charge at C2, transmitted as a field effect, does not appear to have been considered by workers in the field.)

It is evident that the effects of deuterium substitution are interesting. However, the interpretation of secondary isotope effects in solvolysis becomes increasingly complex.[36] Until we have a full understanding of such secondary isotope effects, it would appear highly dangerous to attempt to utilize them to resolve the question of the structure of the 2-norbornyl cation.

11.8. Deactivating Substituents

It was suggested that the introduction of electron-withdrawing substituents might inhibit σ-bridging and yield classical 2-norbornyl cations.[37,38] Indeed, major changes in the *exo*:*endo* rate ratios have been observed in representative systems.[39–41] Representative systems are 7-oxo-2-norbornyl tosylate (**23**),[37] 7,7-dimethoxy-2-norbornyl tosylate (**24**),[38]

exo:*endo* (25°C):   0.17   8

substituted 2-benzonorbornenyl derivatives (**25, 26, 27**),[39] the 6-carbomethoxy-2-norbornyl (**28**),[40] and 1-cyano-2-norbornyl (**29**).[41]

exo:*endo* (77.6°C):   310,000   4600   3.7

[36] V. J. Shiner, Jr., *Isotope Effects in Chemical Reactions*, C. J. Collins and N. S. Bowman, Eds., Van Nostrand–Reinhold, New York, 1970, Chapter 2.
[37] P. G. Gassman and J. L. Marshall, *J. Amer. Chem. Soc.*, **88**, 2822 (1966).
[38] P. G. Gassman and J. L. Marshall, *Tetrahedron Lett.*, 2433 (1968).
[39] H. Tanida, H. Ishitobi, T. Irie, and T. Tsushima, *J. Amer. Chem. Soc.*, **91**, 4512 (1969); H. Tanida, T. Irie, and T. Tshushima, *ibid.*, **92**, 3404 (1970).
[40] G. W. Oxer and D. Wege, *Tetrahedron Lett.*, 457 (1971).
[41] D. Lenoir, *Chem. Ber.*, **108**, 2055 (1975).

28 H		**29** H

exo:*endo* (25°C):   4.4   1.1

It was proposed that these solvolyses proceed through "classical" 2-norbornyl cations and that a low *exo*:*endo* rate ratio was a characteristic of such cations.

Do these results support σ-bridging in the parent system which exhibits a high *exo*:*endo* rate ratio?

Gassman and Macmillan reexamined the solvolysis of the 7-oxygenated norbornyl derivatives, such as **23** and **24**, and concluded that all such derivatives may solvolyze by mechanisms involving participation in the *endo* isomers. They concluded that such substrates could not be used to establish the characteristics of a "classical" 2-norbornyl cation.[42]

Indeed, Gassman and Hornback noted that *syn*- and *anti*-7-chloro-2-norbornyl tosylates solvolyze without participation by the 7-substituent.[43] Yet these derivatives exhibit normal *exo*:*endo* rate ratios in spite of the powerful $-I$ effects of the 7-chloro substituents.

The products obtained in the solvolysis of *exo* and *endo* derivatives of **27**, **28**, and **29** are very different. For example, the acetolysis of *exo*-**27** and *endo*-**27** at 180°C yields the following:

$$\text{(23)}$$

exo	41%	35%	21%
endo	89%	3%	2.2%

It may be recalled that earlier it was pointed out that the acetolysis of 2-norbornyl is borderline for a carbonium ion process (Section 11.2). The introduction of electron-withdrawing substituents evidently transforms the solvolysis into a carbonoid process with solvent participation playing a major role. Clearly these results are not pertinent to the question of the characteristics of a "classical" 2-norbornyl cation.

Pertinent rate data for 2-norbornyl derivatives containing deactivating substituents are summarized in Table 11.5.

[42]P. G. Gassman and J. G. Macmillan, *J. Amer. Chem. Soc.*, **91**, 5527 (1969).
[43]P. G. Gassman and J. M. Hornback, *J. Amer. Chem. Soc.*, **91**, 4280 (1969).

TABLE 11.5. *Effect of Deactivating Substituents on the Rate of Solvolysis of 2-Norbornyl Derivatives*[a]

Compound	X	$10^6 k_1$, sec^{-1} at 25°C	$\Delta H^{\ddagger}$, kcal mol^{-1}	$\Delta S^{\ddagger}$, eu	RR at 25°C	$\dfrac{exo}{endo}$	Reference
	X-OBs	88.2	25.1	7.3	1.00	350	e
	N-OBs	0.252	26.0	−1.5	1.00		
	X-OTs	23.3	21.6	−7.2	1.00	280	19
	N-OTs	0.0828	25.8	−4.4	1.00		
syn-7-Cl	X-OTs	0.0662	26.4	−2.9	0.000284	246	43
	N-OTs	0.00027	29.1	−4.7	0.0032		
anti-7-Cl	X-OTs	0.0429	26.9	−2.0	0.00184	80	43
	N-OTs	0.000538	29.3	−2.7	0.0065		
1-CN[b]	X-OTs	0.000694	26.3	−12.1		1.1	41
	N-OTs	0.000651	29.3	−23.0			
1-COOMe[b]	X-OTs	1.61	22.1	−10.9		12.7	41
	N-OTs	0.126	20.0	−23.0			
6-COOMe	X-OBs	0.0074	29.0	1.7	0.000084	4.4	40
	N-OBs	0.0019	28.9	−1.4	0.00675		
5-Oxo	X-OBs	0.0085	21.1	−24.5	0.000968	43	f
	N-OBs	0.0002	25.8	−15.9	0.000794		
7-Oxo	X-OTs	0.0144	27.1	−3.5	0.00062	0.17	37
	N-OTs	0.0866	24.7	−8.1	1.04		
7-Ethylenedioxy	X-OTs	0.129	28.5	5.6	0.00554	11	42
	N-OTs	0.0120	29.5	4.1	0.145		
7,7-(OMe)$_2$	X-OTs	0.0953	26.9	−0.5	0.0014	8	38
	N-OTs	0.0114[c]	—	—	0.14		
	X-OBs	7.47	24.1	−1.2	0.0847	4,600[d]	39
	N-OBs	0.00051	28.9	−3.9	0.00202		
3'-OMe	X-OBs					310,000[d]	39
	N-OBs	0.00154	28.1	−4.7	0.006		
2',3'-(NO$_2$)$_2$	X-OBs	0.00211	28.8	−1.4	0.000024	3.7[d]	39
	N-OBs	0.0000225	30.0	−6.4	0.000012		

[a] Rate constants, relative rates and *exo*:*endo* rate ratios are at 25°C, unless otherwise indicated.
[b] In 60% ethanol.
[c] Estimated value, see Reference 1.
[d] At 77.6°C.
[e] S. Winstein *et al.*, *J. Amer. Chem. Soc.*, **74**, 1127 (1952).
[f] J. C. Greever and D. E. Gwynn, *Tetrahedron Lett.*, 813 (1969).

Finally, it was shown that the rates of acetolysis of the 7-substituted *exo*-norbornyl tosylates are well correlated by σ^*.[44] It was concluded that if there was no participation in the solvolysis of 7-keto-*exo*-norbornyl tosylate, then the maximum contribution of σ-participation to the acetolysis of *exo*-norbornyl tosylate itself is a factor less than 5.

11.9. Structural Modifications

Numerous structural modifications have been made in an attempt to deduce whether the results are consistent with and therefore support the σ-bridged structure in the parent system. Thus the low *exo*:*endo* rate ratio in *exo*-5,6-trimethylene-2-norbornyl (30)[45] and in *exo*-4,5-trimethylene-2-norbornyl (31)[46] are attributed to steric difficulties in achieving σ-bridging.[1] Consequently, the high *exo*:*endo* rate ratio in 2-norbornyl itself is attributed to σ-bridging.

	30 H	31 H
exo:*endo* (25°C):	9.5	8.5

On this basis it is puzzling that *exo*-5,6- (32) and *endo*-5,6-methylene-2-norbornyl (33) brosylates exhibit high *exo*:*endo* rate ratios.[47] The same argument used to account for the behavior of 30[1] would also predict an equally low *exo*:*endo* rate ratio for 32.

	32 H	33 H
exo:*endo* (25°C):	1150	470

The acetolysis of *exo*-5,6-*o*-phenylene-2-norbornyl tosylates also yields exceptionally low *exo*:*endo* rate ratios (34).[48] On the other hand, the *endo*-5,6-*o*-phenylene-2-norbornyl tosylates (35) yield a high *exo*:*endo*

[44]P. G. Gassman, J. L. Marshall, J. G. Macmillan, and J. M. Hornback, *J. Amer. Chem. Soc.,* **91**, 4282 (1969).
[45]K. Takeuchi, T. Oshika, and Y. Koga, *Bull. Chem. Soc. Japan,* **38**, 1318 (1965).
[46]E. J. Corey and R. S. Glass, *J. Amer. Chem. Soc.,* **89**, 2600 (1967).
[47]K. Wiberg and G. R. Wenzinger, *J. Org. Chem.,* **30**, 2278 (1965).
[48]R. Baker and J. Hudec, *Chem. Commun.,* 929 (1967).

rate ratio.[48] An examination of the rate data (Table 11.6) reveals that the high *exo*:*endo* rate ratio in **35** arises primarily from a remarkably slow rate for the *endo*-2-tosylate and not from an enhanced rate for the *exo*-2 tosylate. Thus even in this system the high *exo*:*endo* rate ratio must be attributed primarily to steric hindrance to ionization in the *endo* isomer.

	34	H	**35**	H
exo:*endo* (25°C):		1.1		1270

TABLE 11.6. *Effect of Structure Modifications on the Rate of Solvolysis of 2-Norbornyl Derivatives*

Compound	X	$10^6 k_1$, sec^{-1} at 25°C	$\Delta H^{\ddagger}$, kcal mol^{-1}	$\Delta S^{\ddagger}$, eu	RR at 25°C	$\dfrac{exo}{endo}$	Refer-ence
	X-OBs	88.2	25.1	7.3	1.00	280	a
	N-OBs	0.252	26.0	−1.5	1.00		
	X-OTs	23.3	21.6	−7.2	1.00	350	19
	N-OTs	0.0828	25.8	−4.4	1.00		
	X-OTs	0.323	26.4	0.3	0.0138	9.5	45
	N-OTs	0.034	26.4	−4.1	0.41		
	X-OTs	1.55	26.1	2.3	0.066	186	45
	N-OTs	0.00829	26.8	−5.5	0.10		
	X-OTs	0.283	25.6	−2.6	0.012	8.5	46
	N-OTs	0.0328	26.7	−3.1	0.4		
	X-OBs	13.6	24.2	1.3	0.154	1150	47
	N-OBs	0.0118	27.5	−2.5	0.046		

TABLE 11.6 (continued)

Compound	X	$10^6 k_1$, sec^{-1} at 25°C	$\Delta H^{\ddagger}$, kcal mol^{-1}	$\Delta S^{\ddagger}$, eu	RR at 25°C	*exo/endo*	Reference
	X-OBs N-OBs	6.8 0.0145	26.8	−4.5	0.077 0.057	470	47
	X-OTs N-OTs	0.00575 0.00528	29.1 27.3	1.5 −4.8	0.00027 0.063	1.1	48
	X-OTs N-OTs	0.0763 0.00006	24.1 30.8	−10.2 −2.1	0.0033 0.00072	1270	48
	X-OTs N-OTs	0.0071 0.0233	29.2 28.5	2.1 2.3	0.000304 0.25	0.35	b
	X-OBs N-OBs	0.836 0.0376	26.2 27.6	1.4 0.1	0.0095 0.149	22	c
	X-OBs N-OBs	8.54 2.30			0.097 9.2	3.7	49
	X-OBs N-OBs	39.1 57.0			0.443 226	0.68	49

[a] S. Winstein *et al.*, *J. Amer. Chem. Soc.*, **74**, 1127 (1952).
[b] I. Rothberg *et al.*, *J. Amer. Chem. Soc.*, **92**, 2570 (1970).
[c] L. de Vries and S. Winstein, *J. Amer. Chem. Soc.*, **82**, 5363 (1960).

In these systems it is necessary to guard against a transition from a carbonium ion process to a carbonoid process. For example, the acetolysis of the fused norbornyl brosylates **36** exhibits an *exo* : *endo* rate ratio of 0.69.[49]

36

exo : *endo* (25°C):   0.69

In spite of the low *exo* : *endo* rate ratio, the authors utilize a σ-bridged intermediate to interpret their results. However, the products appear to be in better accord with a carbonoid process (24).

(24)

exo-**36** $\xrightarrow[\substack{HOAc\\NaOAc}]{50°C}$ 65.3%	27.3%	7.4%	
endo-**36** $\xrightarrow[\substack{HOAc\\NaOAc}]{50°C}$ 96.3%	3.1%	0.6%	

Pertinent data are summarized in Table 11.6.

A great deal of effort has been expended on the study of these modified 2-norbornyl systems. Regrettably, in almost no cases have the authors established that they are dealing with a carbonium ion process with a rate-product relationship expressable in a Goering–Schewene diagram. Until such results are available, we are not in position to achieve a true understanding of these structural effects and their pertinence to the question of σ-bridging in the 2-norbornyl cation.

11.10. Conclusion

Simple tests for the proposed charge delocalization to the 6-position have uniformly yielded negative answers. Standard tests for the presence of resonance stabilization at the 2-position have yielded negative answers. Tests for charge delocalization to the 1-position in systems not undergoing rearrangement to a more stable structure have yielded negative answers.

[49]R. K. Howe, P. Carter, and S. Winstein, *J. Org. Chem.*, **37**, 1473 (1972).

It would appear that the more qualitative the data and the less understood the phenomena (Sections 11.7–11.9) the more they have been applied as arguments to support the σ-bridged structure for 2-norbornyl. Such data and phenomena must be better understood if they are to contribute to solving the problem.

Comments

The modest change in *exo*:*endo* solvolytic rate ratios in going from secondary to tertiary 2-norbornyl derivatives (1) is one of Brown's strongest arguments. However, it is my view that the similarity of such ratios, when such similarities are found, are only fortuitous; I believe the underlying causes to be different.

Brown has stressed not only the similarity of *exo*:*endo* ratios, but also that of the free-energy diagrams (e.g., Figures 11.1 and 11.2). However, the free-energy diagram for the *t*-butyl-2-norbornyl system[4] (Figure 8.9) differs significantly; the *endo*-OPNB (**37**) is more stable than its epimer (**38**), and ground state and transition state effects are equally important. Only by luck does **38**:**37** seem to fit in series (1).

There are many instances where the *exo*:*endo* ratios do not remain constant. Compare Figures 8.6, 8.7, and 8.8. The ground-state energy differences in these tertiary esters are essentially the same; the large change in the *exo*:*endo* ratios (from 6.1 to 3,630,000) are entirely due to transition-state energy variations. As I pointed out in my commentaries to Chapter 8, the corresponding secondary esters all show similar *exo*:*endo* ratios (206 to 4100). Tertiary and secondary systems are behaving in fundamentally different ways.

Some of the most damaging evidence against his own position has been collected by Brown himself.[50] The tertiary *exo*-5,6-trimethyl-enenorbornane derivatives behave just like their 2-norbornyl counterparts as far as *exo*- and *endo*-rates, and *exo*:*endo* ratios are concerned (25).

| exo:endo (25°C): | 885 | 420 | 127 | 118 |

(25)

There is no evidence that the extra five-membered ring is introducing any steric complication and, after all, such steric effects should be largest for

---

[50]H. C. Brown, D. L. Vander Jagt, P. v. R. Schleyer, R. C. Fort, Jr., and W. E. Watts, *J. Amer. Chem. Soc.*, **91**, 6848 (1969).

the tertiary derivatives. No dipolar effects are anticipated, and the five-membered ring substituent extends away from the 2-*exo* and the 2-*endo*-positions and leaving groups. Force-field calculations show no unusual strains, twisting effects, or distortion effects introduced by the 5,6-trimethylene substituent.[51] On the basis of classical carbocation theory, one would expect the secondary esters **30** to exhibit the "usual" large *exo*: *endo* ratio. *Instead, a sharp decrease to 9.5 is found.* Despite the "softness' of classical carbocation theory, no satisfactory rationalization for this decrease, and that of the similar cases presented in Table 11.6, has been offered. Nonclassical theory offers a simple and convincing explanation.[1] Systems like **32**, **33**, and **35** clearly are more complex than **30** or **31** because of *endo* crowding or because of the strain or electrical effects introduced.

I agree with both Brown and Sargent[1] that the tertiary *exo*: *endo* ratios are largely steric in origin. Steric hindrance to ionization of the *endo* derivatives is the most plausible explanation for the high ratios observed. The origin of high *exo*: *endo* ratios in secondary norbornyl derivatives must be electronic, because these ratios generally respond in a different way to the effects of substituents (also see comments, Chapter 10).

Deactivating substituents also can reduce secondary *exo*: *endo* ratios to low values. Of the many cases which have been examined (Table 11.5), I am most impressed by the results for **28**; the substituent is well removed from the reaction site. Even so, dipolar effects may be significant, and it would be desirable to provide calibration by studying the corresponding tertiary derivatives. The reinvestigation of **28** in limiting solvents would overcome Brown's objection about the products not being the same from *endo* and *exo* starting materials. I doubt if this would alter the *exo*: *endo* ratios very much, or the conclusion that the 6-carbomethoxy group in **28** is suppressing σ-participation.

Winstein realized quite early that the bridging carbon (C6 in the nonclassical norbornyl cation) remains essentially sp^3 hybridized and bears little charge.[52] *Ab initio* calculations have now verified this conclusion.[53] π-Donating substituents like CH$_3$ or CH$_3$O at C6 thus are not able to promote participation. Our work in this area[18–20] was undertaken to verify these ideas.

Coates' cation is now accepted to be a bona fide nonclassical σ-bridged ion (see comments, Chapter 14).[54,55] However, the substituent effects exhi-

[51]E. M. Engler and S. Godleski, unpublished calculations. See E. M. Engler, J. D. Andose, and P. v. R. Schleyer, *J. Amer. Chem. Soc.*, **95**, 8005 (1973).

[52]R. J. Piccolini and S. Winstein, *Tetrahedron Supplement*, **19**, 423 (1963); R. Howe, E. C. Friedrich, and S. Winstein, *J. Amer. Chem. Soc.*, **87**, 379 (1965).

[53]L. Radom, J. A. Pople, V. Buss, and P. v. R. Schleyer, *J. Amer. Chem. Soc.*, **94**, 311 (1972).

[54]R. M. Coates and E. R. Fretz, *J. Amer. Chem. Soc.*, **99**, 297 (1977).

[55]H. C. Brown and M. Ravindranathan, *J. Amer. Chem. Soc.*, **99**, 299 (1977).

bited by this system (26) are even more modest than those found (Table 11.3) for 1-methyl- and 1-phenyl-2-*exo*-norbornyl solvolyses. Competing σ-withdrawing and π-donating effects may be involved. Whatever the explanation, such behavior appears to be characteristic of nonclassical systems.[56]

RR (25°C):   1.0   4.8   0.04

$$(26)$$

Classical carbocation theory does no better in explaining such behavior in unsymmetrical systems. The large acceleration produced by the 1-$CH_2Sn(CH_3)_3$ substituent in **11**, 6×10^5, can be rationalized by arguing that σ-participation of the C6,1 electrons occurs during ionization, and that the first intermediate is not a bridged ion, but a rearranged highly stabilized classical 2-substituted norbornyl cation. If this is the case, why is a 1-phenyl or 1-anisyl group so ineffective (Table 11.3)?

The important feature of the isotope effect data summarized in (20) is not the magnitude of the effects, but rather the different behavior shown by the secondary (**16**) and tertiary (**17,18**) systems. In **16**, k_H/k_D is much smaller for the *exo* derivative than for the *endo*, but for both **17** and **18**, the *exo* and *endo* values are the same. The conclusion that participation selectively reduces k_H/k_D for *exo*-**16** is supported by parallel results in the analogous anchimerically assisted benzonorbornenyl derivatives (27).[57]

$$(27)$$

| k_H/k_D (HOAc): | 1.05 (25°C) | 1.28 (100°C) |
| (HOOCH): | 1.04 (25°C) | 1.24 (100°C) |

A letter I wrote to Professor Brown[58] clarifies some of the points concerning the mechanism of 2-*exo*- and 2-*endo*-norbornyl solvolysis.

---

[56]See the compilation of data in W. F. Sliwinski, T. M. Su, and P. v. R. Schleyer, *J. Amer. Chem. Soc.*, **94**, 133 ((1972).
[57]H. Tanida and T. Tsushima, *J. Amer. Chem. Soc.*, **93**, 3011 (1971).
[58]P. v. R. Schleyer to H. C. Brown, Jan. 12, 1975.

"It is obvious that k_c and k_Δ-type substrates should plot against one another, since both depend on the 'ionizing power' of the solvent and not on solvent nucleophilicity. However, as Winstein found, many substrates of the k_s-type plot satisfactorily against Y constants, provided the degree of nucleophilic assistance is relatively small. One should not be surprised to find 2-*endo*-norbornyl tosylate in this category. There is no significant statistical improvement in using the full four parameter equation, $\log (k/k_0) = mY + lN$.[59] In short, this kind of plot is too insensitive to reveal the existence of relatively small amounts of nucleophilic solvent assistance.

The basis upon which one can conclude that 2-*endo*-norbornyl enjoys a modest amount of such solvent assistance is based on the slope of your plot against 2-adamantyl, 0.664. This suggests a greater degree of charge dispersion in the transition state than is enjoyed by 2-adamantyl. Of course, the partially bridged transition state in 2-*exo* solvolysis would also lead to a greater amount of charge dispersion than is present in 2-adamantyl, and the slope you find, 0.786, indicates this. Other k_Δ substrates (phenonium ion cases afford unambiguous examples) afford similarly reduced m-values. In addition, the production of some optically active 2-*exo*-norbornyl product from the solvolysis of optically active 2-*endo* substrates strongly suggest the intervention of a weak k_s mechanism. Leakage to the bridged ion predominates, but some of the nucleophilically solvated intermediate ion pair collapses directly to product."

The interpretation of 2-*endo*-norbornyl solvolysis as being solvent assisted is supported by a k_s/k_c ratio in HOAc of 30^{13} (vs. 93^{13} or 105^{59} for cyclopentyl, 31^{13} for *trans*-2-methylcyclopentyl, and 28^{59} for cyclohexyl) (see Table 11.1). In addition, 3-*exo* substituents, which should block solvent access to the rear, reduce the 2-*endo* solvolysis rates appreciably (Table 11.4) (28).[26]

	OTs	OTs	OTs (28)
RR (25°C):	1.0	1/260	1/11.5

Activation volumes seem to distinguish between anchimerically assisted and solvent assisted solvolyses.[60] On this basis, 2-*exo*-norbornyl

[59] F. L. Shadt, T. W. Bentley, and P. v. R. Schleyer, *J. Amer. Chem. Soc.*, **98**, 7667 (1976). Also see T. W. Bentley and P. v. R. Schleyer, *ibid.*, **98**, 7658 (1976).
[60] W. J. le Noble, B. L. Yates, and A. W. Scaplehorn, *J. Amer. Chem. Soc.*, **89**, 3751 (1967); W. J. le Noble and A. Shurpik, *J. Org. Chem.*, **35**, 3588 (1970).

falls into the former category, while 2-*endo*-norbornyl is normal. There seems to be a high degree of consistency in the results obtained by this method.[61]

I quite agree that "we are dealing with a borderline situation which might easily be shifted by substituents or structure modifications to a process in which the products from *exo* and *endo* are clearly different, no longer attributable to a common intermediate". Compounds **27**, **28**, **29**, and **36** afford examples of such modified behavior (23), (24), but acetolysis of typical epimeric secondary tosylates also give different product mixtures.[62] The point about norbornyl, which so impressed Winstein, was that it behaved so differently from other secondary tosylate systems examined in the early days of solvolysis. As I have pointed out in Comments 6, high epimeric rate ratios are unknown in any secondary system not exhibiting the characteristics of anchimeric assistance.

We now realize the choice of acetic acid as the "standard" solvolysis solvent was unfortunate. Work on substrates like **27–29**, and **36** could now be repeated in more limiting solvents, but I doubt if the *exo*: *endo* ratios or the conclusions would be altered significantly.

[*Note added in proof*] Maskill[63] has carefully redetermined the secondary deuterium isotope effects during 2-*exo* and 2-*endo*-norbornyl-2-*d* solvolyses. By incorporating a second deuterium at C-1, problems involving scrambling due to internal return are avoided. His conclusions are remarkably unambiguous, "These results are in accordance with a mechanism involving bridging at the transition state for ionization which leads to the nonclassical norbornyl cation. The results are incompatible with a mechanism involving direct ionization to interconverting classical cations."

[*Note added in proof* (HCB)] Two of Paul Schleyer's former coworkers have made recent contributions. Fărcaşiu has analyzed homoretroene eliminations in organometallic derivatives and has pointed out that the fast conversion of **11** to **12** could be such a reaction.[64] Harris has applied his test for k_c vs k_s processes[65] and concluded that solvolysis of *endo*-norbornyl tosylate is k_c.[66]

[61]W. J. le Noble, E. H. White, and P. M. Dzadzic, *J. Amer. Chem. Soc.*, **98**, 4020 (1976).

[62]Recent reviews: J. M. Harris, *Progr. Phys. Org. Chem.*, **11**, 89 (1974); J. M. Harris and C. Wamser, *Fundamentals of Organic Reaction Mechanisms*, Wiley, New York, 1976, Chapter 4; T. W. Bentley and P. v. R. Schleyer, *Adv. Phys. Org. Chem.*, **14** (1977) in press.

[63]H. Maskill, *J. Amer. Chem. Soc.*, **98**, 8482 (1976).

[64]D. Fărcaşiu, *Tetrahedron Lett.*, 595 (1977).

[65]J. M. Harris, D. J. Raber, W. C. Neal, Jr., and M. D. Dukes, *Tetrahedron Lett.*, 2331 (1974).

[66]J. M. Harris, D. L. Mount, and D. J. Raber, manuscript submitted.

12

Capture of Unsymmetrical 2-Norbornyl Cations

12.1. Introduction

The direct tests with appropriate substituents for charge delocalization from the 2-[1] to the 1-[2] and 6-[3,4] positions in solvolytic reactions of 2-norbornyl derivatives all failed to confirm such charge delocalization (Chapter 11). It was then suggested that "there are good reasons to expect carbon bridging to lag behind C–X ionization at the transition state."[5]

If this position is accepted, there is little point in attempting either to prove or to disprove the existence of σ-participation through a study of the rates of solvolysis of 2-norbornyl derivatives. Accordingly, we turned our attention to the question of whether there are carbonium ion reactions in which the existence of the 2-norbornyl cation in an unsymmetrical state can be demonstrated.

12.2. Deamination of 2-Norbornylamine

It was originally reported that the deamination of (+)-endo-norbornylamine in acetic acid affords exo-norbornyl acetate with ca. 23% retention of optical purity.[6] The authors did not conclude that they had

---

[1]H. C. Brown and M.-H. Rei, J. Amer. Chem. Soc., 86, 5008 (1964).
[2]P. v. R. Schleyer and D. C. Kleinfelter, 138th Meeting of the American Chemical Society, New York, 1960, Abstracts, p. 43P. Data in J. A. Berson, Molecular Rearrangements, P. de Mayo, Ed., Interscience, New York, 1963, p. 182.
[3]P. v. R. Schleyer, M. M. Donaldson, and W. E. Watts, J. Amer. Chem. Soc., 87, 375 (1965).
[4]P. v. R. Schleyer, P. J. Stang, and D. J. Raber, J. Amer. Chem. Soc., 92, 4725 (1970).
[5]S. Winstein, J. Amer. Chem. Soc., 87, 381 (1965).
[6]J. A. Berson and D. A. Ben-Efraim, J. Amer. Chem. Soc., 81, 4094 (1959).

successfully trapped the 2-norbornyl cation in its classical, unsymmetrical form. Instead, they suggested that the formation of the optically active *exo*-norbornyl acetate must be derived "by direct displacement of the solvent on the diazonium ion."[6]

Corey and his coworkers[7] felt that this hypothesis was not consistent with the much smaller displacement by solvent (7%) observed in the solvolysis of the brosylate.[8] Accordingly, they examined the deamination of optically active *exo*-norbornyl amine (**1**). The product was 96% *exo*-norbornyl acetate (**2**) with about 4% *endo*-. There was realized 15% retention of optical activity (**1**).

$$\underset{\substack{\mathbf{1}\\(-)}}{\text{\includegraphics{NH}_2 \cdot \text{HCl}}} \xrightarrow[\text{NaNO}_2]{\text{HOAc, 20°C}} \underset{\substack{\mathbf{2}\\(-)\,15\%\,\text{opt. pur.}}}{\text{OAc}} \tag{1}$$

Treatment of *endo*-norbornylamine-$2d_1$ (**3**) under the same conditions yielded *exo*-norbornyl acetate (**4**) containing 58% of the deuterium at position-2 (2). This result corresponds to the 16% retention of optical activity observed in the stereochemical experiment utilized for the *exo* amine.

$$\underset{\substack{\mathbf{3}\\ \text{NH}_2\cdot\text{HCl}}}{\text{D}} \xrightarrow[\text{NaNO}_2]{\text{HOAc, 20°C}} \underset{\substack{\mathbf{4}\\ \text{D}\\ 58\%}}{\text{OAc}} + \underset{\substack{\text{D}\\ 42\%\\ \text{(plus other}\\ \text{possible isomers)}}}{\text{OAc}} \tag{2}$$

The authors conclude that both *exo* and *endo* amines are converted essentially completely to a classical 2-norbornyl cation which serves as a common intermediate for both rearranged and unrearranged products.

In a subsequent, more detailed study Berson and Remanick confirmed the major features of this study.[9] Thus optically active *exo*-norbornylamine yielded 87% 2-norbornyl acetate (98% *exo*-, 2% *endo*). The *exo* acetate revealed $11 \pm 2\%$ activity. The *endo* amine yielded 80% 2-norbornyl acetate (95.3% *exo*- and 5.3% *endo*). The *exo* acetate was $18 \pm 0.6\%$ active.

However, the authors stress certain differences. Thus, the *endo* amine yielded approximately twice as much *endo* acetate as did the *exo* amine. The optical activity of the *exo* acetate from the *endo* amine is greater (18%) than

[7]E. J. Corey, J. Casanova, Jr., P. A. Vatakencherry, and R. Winter, *J. Amer. Chem. Soc.*, **85**, 169 (1963).
[8]S. Winstein and D. Trifan, *J. Amer. Chem. Soc.*, **74**, 1147, 1154 (1952).
[9]J. A. Berson and A. Remanick, *J. Amer. Chem. Soc.*, **86**, 1749 (1964).

that from the *exo* amine (11%). Finally, little racemization was observed in the small amount of *endo* acetate (optical purity: $85 \pm 12\%$) derived from the *endo* amine. Consequently, these authors conclude that the products cannot be derived solely from Wagner–Meerwein interconversion of two enantiomeric 2-norbornyl classical ions. They prefer a mechanism in which the classical 2-norbornyl cation is first formed and is captured competitively with its conversion to the bridged species.

Irrespective of what would appear to be minor differences in these interpretations, both groups agree that in this reaction they have generated the enantiomeric 2-norbornyl classical ion and have captured it before it has completely racemized, either by rapid equilibration or by conversion to a symmetrical bridged species.

The success in generating and trapping of the classical 2-norbornyl cation in this reaction is attributed to the fact that the highly exothermic loss of nitrogen from the diazonium ion requires only a small activation energy, approximately 5 kcal mol^{-1}, with little need for σ-participation.[10] It is argued that the more endothermic reaction, solvolysis, with an activation energy of $\sim$25 kcal mol^{-1}, utilizes such σ-participation.

An alternative interpretation is that the *exo*-norbornyl derivatives solvolyze to tight ion pairs.[11] Such ion pairs in bicyclic systems, where backside capture of solvent is difficult for steric reasons, may have a relatively long life in which many equilibrations can occur before solvent can penetrate the rapidly equilibrating ion pair and be captured. In terms of this picture, deamination, with its rapid loss of nitrogen, provides a much looser ion pair than does solvolysis.

Irrespective of the final explanation, one should not lose sight of the fact that these deamination reactions provide unambiguous evidence for the formation and capture of the 2-norbornyl cation in the optically active, classical form.

12.3. Attempted Trapping of Unsymmetrical Cations in Solvolytic Processes

Corey and his coworkers advanced an interesting new idea.[7] By utilizing *m*-carboxybenzenesulfonate as the leaving group, they proposed to trap the 2-norbornyl cation prior to full equilibration. They prepared *p*-nitrophenyl-(*m*-chlorosulfonyl)benzoate. This was used to convert optically active 2-norborneol into the arenesulfonate (**5**). Unfortunately, this was an

[10]G. D. Sargent, *Carbonium Ions*, G. A. Olah and P. v. R. Schleyer, Eds., Vol. III, Wiley–Interscience, New York, 1972, Chapter 24.
[11]R. A. Sneen, *Accounts Chem. Res.*, **6**, 46 (1973).

oil which was not characterized. This crude product was then solvolyzed in
tert-butyl alcohol and in aqueous tetrahydrofuran, with sufficient sodium
hydroxide added to saponify the *p*-nitrophenyl ester. In theory the reaction
should have proceeded to yield the ion pair (**6**) rapidly converted into the
benzoate ester (**7**) (3).

(3)

Not only was the initial ester not characterized, but none of the reaction
products were isolated and characterized. Perhaps even more serious, the
technique was not tested in any system known to equilibrate rapidly. All that
can be said at the present time is that the 2-norborneol, isolated in poor yield
from such intermediates as were formed in the solvolyses, exhibited no
activity.

In a similar study, Smith and Petrovich examined the solvolysis of
optically active *exo*-norbornyl *p*-trifluoromethylthiobenzoate (**8**) in acetic
acid and aqueous ethanol.[12] They postulated that capture of the first formed
ion pair (**9**) by the sulfur atom (**10**) would be very fast and provide hope of
capturing it in optically active form if the cation is formed in the unbridged
form (4).

(4)

The acetolysis products were 73% *exo*-norbornyl acetate and 16%
exo-norbornyl *p*-trifluoromethylthiobenzoate (**10**). The thiobenzoate **10**
was inactive, so Smith and Petrovich concluded that the 2-norbornyl
cation must be symmetrical toward ion-pair return by sulfur.

It is evident that their own results disprove their original postulate. If
ion-pair return to sulfur is so fast that one can hope to capture the
equilibrating pair of cations in its unsymmetrical form, then it must necessarily be much faster than the capture of solvent by the ion pair. However, the

[12]S. G. Smith and J. P. Petrovich, *J. Org. Chem.*, **30**, 2882 (1965).

predominant product is the acetate (73%), not the *p*-trifluoro-methylthiobenzoate (16%). Clearly, ion-pair return to sulfur is not faster, but actually slower than capture of the intermediate by solvent.

This is an interesting, promising approach to the problem. Regrettably, it is not possible to say at this time that either the technique or the possibility of capturing the 2-norbornyl cation by the application of this technique have yet received a fair trial.

12.4. Capture of Unsymmetrical Cations in Solvolytic Processes

The acetolysis of tagged 2-(Δ^3-cyclopentenyl)ethyl-2-^{14}C *p*-nitrobenzenesulfonate (**11**) was studied by Lee and Lam.[13] They characterized the various isomeric tagged species formed (5). Three *exo*-norbornyl acetates were identified: **12**, 37%; **13**, 38%; **14**, 25%.

(5)

It was pointed out by Collins and Lietzke that the formation of **13** and **14** in unequal amounts is inconsistent with the formulation of the pair of cations **15** and **16**, as a resonance hybrid.[14] Such a resonance hybrid would require the formation of equal amounts of the two acetates, **13** and **14**. On the other hand, the formation of an excess of **13** over **14** is consistent with the formation of the classical ion **15**, which is captured competitively with its fast Wagner–Meerwein conversion into **16**.

[13] C. C. Lee and L. K. M. Lam, *J. Amer. Chem. Soc.*, **88**, 2834 (1966).
[14] C. J. Collins and M. H. Lietzke, *J. Amer. Chem. Soc.*, **89**, 6565 (1967).

Collins and Lietzke pointed out that with $k_s/k_H = 0.022$ and $k_s/k_{W-M} = 0.33$, they could reproduce the ^{14}C-product distribution realized by Lee and Lam. These rate ratios are different from those observed in the solvolysis of 2-norbornyl derivatives. They indicate too low a rate for the Wagner–Meerwein interconversion (k_{W-M}) of the two classical cations and too high a rate for the 6,2-hydride shift (k_H) relative to the rate of solvent capture (k_s) to account for the high degree of racemization and lower amount of hydride migration in normal solvolyses. However, these differences may again be associated with the position of the anion.[15,16] The formation of the 2-norbornyl cation by the π-route (5) places the anion in a position where it is not intimately associated with the cationic center. This could well have the observed consequences.

This result has been confirmed by a study of the capture of the hydride-shifted 2-norbornyl cation from the solvolysis of *exo*-norbornyl-4-^{13}C tosylate (17) in acetic acid (6).[17]

Again, if the ions produced in the 3:2 hydrogen shift, 18 and 19, were a resonance hybrid, the yield of the two acetates, 20 and 21, would be identical. However, once again there is realized a larger yield of that ion which would be the first formed member of a Wagner–Meerwein equilibrating ion.

(6)

20 1.52%   **21** 0.52%

The simplest explanation of these results is that a rapidly equilibrating Wagner–Meerwein pair of 2-norbornyl cations, formed without a tightly associated anion, is being captured prior to complete equilibration.

[15]P. D. Bartlett, *Ann.*, **653**, 45 (1962).
[16]P. D. Bartlett and G. D. Sargent, *J. Amer. Chem. Soc.*, **87**, 1297 (1965).
[17]C. J. Collins and C. E. Harding, *Liebig's Ann.*, **745**, 124 (1971).

12.5. Additions to 2,3-Dideuterionorbornene

The results discussed earlier in this chapter are consistent with the proposal that it is the formation of tight ion pairs in solvolytic processes which resists capture of the 2-norbornyl cation before it has fully equilibrated. Another means of avoiding the formation of such tight ion pairs is to form the 2-norbornyl cation by means of a protonation process. In order to establish whether the reaction is proceeding through a symmetrical or an unsymmetrical intermediate, a tag is required. The use of 2,3-dideuterionorbornene provides an internal tag.[18] Alternatively, as discussed later, the use of deuterated acids provides an external tag.

Addition of hydrogen chloride to 2,3-dideuterionorbornene (**22**) gives an excess of the isomer **23** with retained structure (7). This result is

$$\tag{7}$$

consistent with a mechanism involving formation of the tagged 2-norbornyl cation (**25**) which is captured prior to full equilibration (**25 ⇌ 26**) (8).

$$\tag{8}$$

Reaction of 2,3-dideuterionorbornene with aqueous hydrobromic acid again gives an excess of the isomer **27** with retained structure (9).

$$\tag{9}$$

[18]J. K. Stille and R. D. Hughes, *J. Org. Chem.*, **36**, 340 (1971).

Finally, the acid-catalyzed reaction of 2,3-dideuterionorbornene with phenol produces the O-alkyl isomer **29 – 30** with an excess of the retained structure **29** (10).

$$\text{(10)}$$

47%   39%

All of these results are consistent with a trapping of the unsymmetrical classical 2-norbornyl cation before it has become fully equilibrated.

12.6. Addition of Deuterium Chloride to Norbornene

The addition of hydrogen chloride to norbornene (**31**) in ethyl ether, methylene chloride, or pentane proceeds rapidly at −78°C to yield *exo*-norbornyl chloride (**32**) in an isomeric purity of at least 99.5% (11).[19]

$$\text{(11)}$$

2-Methylenenorbornane (**33**) adds hydrogen chloride rapidly to give the tertiary chloride **34** as the initial product. This is rapidly converted into the secondary chloride **35** on further treatment with hydrogen chloride (12).

$$\text{(12)}$$

Similarly, 2-methylnorbornene (**36**) is converted initially into the same tertiary chloride **37** (13).

$$\text{(13)}$$

[19]H. C. Brown and K.-T. Liu, *J. Amer. Chem. Soc.*, **97**, 600 (1975).

Remarkably, 1-methylnorbornene (**38**) adds hydrogen chloride to give a mixture of 45% of 4-methyl-*exo*-norbornyl chloride (**39**) and 55% of the tertiary chloride **40** (14).

$$(14)$$

Finally, the addition of hydrogen chloride to 7,7-dimethylnorbornene (**41**) yields 90% of 7,7-dimethyl-*exo*-norbornyl chloride (**42**) and 10% of the Wagner–Meerwein shifted product, 3,3-dimethyl-*exo*-norbornyl chloride (**43**) (15).

$$(15)$$

The strong directive influence leading to the exclusive formation of the tertiary chlorides from **33** and **36** are characteristic of carbonium ion processes. The small discrimination between protonation at C2 and C3 of **38** reveals that the transition state for the protonation reaction involves much smaller development of positive charge than does that for solvolysis, where a 1-methyl group is strongly activating (Section 11.6). However, following protonation, the intermediate must be the carbonium ion in which we are interested. Molecular addition of hydrogen chloride[20] cannot be significant because such an addition would be expected to give the two secondary chlorides from **38**. Both secondary chlorides are stable to the reaction conditions. Moreover, molecular addition to **41** would be expected to give the *endo* isomer predominantly.[21] Consequently, it appears clear that these additions of hydrogen chloride proceed *via* proton transfer from the hydrogen chloride to the olefin with the formation of the 2-norbornyl cation as an intermediate.

The critical experiment involves the addition of deuterium chloride to norbornene. A nonclassical ion intermediate (**44**) requires that the tag be equally distributed between the *exo*-3 (**47**) and *syn*-7 (**48**) positions (16).

$$(16)$$

+hydride-shifted products

[20]R. C. Fahey, *Top. Stereochem.*, **3**, 253 (1969).

[21]H. C. Brown, J. H. Kawakami, and K.-T. Liu, *J. Amer. Chem. Soc.*, **95**, 2209 (1973).

TABLE 12.1. Analyses of the Deuterium Distribution in exo-Norbornyl-d-Chloride Produced in the Addition of Deuterium Chloride to Norbornene

	Yield, %		
Method of analysis			
220-MHz spectrum of **32-d₁**	57	41	2
100-MHz spectrum of **32-d₁**	58	38	4
60-MHz spectrum of **32-d₁**	61	32	7
60-MHz spectrum of olefin from **32-d₁**	59	36	5

The results of a variety of analyses reveal the formation of 57–61% of *exo*-3-*d* (**47**), 32–41% *syn*-7-*d* (**48**), with 2–7% of hydride-shifted material (Table 12.1).

These results cannot be accounted for in terms of the sole formation of a nonclassical intermediate (16). They could be accounted for in terms of the concurrent formation of symmetrical σ-bridged cations (**44**) and rapidly equilibrating classical 2-norbornyl cations (**45⇌46**). However, there appears to be no advantage in introducing this additional complication. The results are most simply accounted for in terms of the formation of a rapidly equilibrating pair of classical cations (**45 ⇌ 46**) which are captured short of full equilibration (17).

(17)

12.7. Addition of Deuteriotrifluoroacetic Acid to Norbornene

The study was extended to trifluoroacetic acid.[22] It possesses exceptionally weak nucleophilic characteristics and has been utilized extensively as a

[22]H. C. Brown and K.-T. Liu, *J. Amer. Chem. Soc.*, **97**, 2469 (1975).

medium for the solvolysis of organic tosylates where it is desirable to minimize such nucleophilic contributions.[23,24] Moreover, the addition of trifluoroacetic acid to olefins shows the characteristics of a typical carbonium ion reaction.[25] Indeed, the data reveal that the addition of trifluoroacetic acid to representative olefins (19) exhibits inductive effects similar to those observed in the trifluoroacetolysis of the related tosylates (18).[26] Accordingly, a study was made of the addition of trifluoroacetic acid[22] and deuteriotrifluoroacetic acid[22] to norbornene and related olefins to test the generality of the results with hydrogen chloride.[19]

$$\underset{\underset{X}{\mid}\quad\underset{OTs}{\mid}}{CH_2CH_2CHCH_3} \xrightarrow{CF_3CO_2H} \underset{\underset{X}{\mid}\quad\underset{O_2CCF_3}{\mid}}{CH_2CH_2CHCH_3} \tag{18}$$

RR: (X = H)/(X = Cl) is 329/1

$$\underset{\underset{X}{\mid}}{CH_2CH_2CH=CH_2} \xrightarrow{CF_3CO_2H} \underset{\underset{X}{\mid}\quad\underset{O_2CCF_3}{\mid}}{CH_2CH_2CHCH_3} \tag{19}$$

RR: (X = H)/(X = Cl) is 421/1

Here also Markovnikov addition to 2-methylenenorbornane was observed (33). No *endo* product was observed with 7,7-dimethylnorbornene (41), so that cyclic addition[20] is evidently not a factor. The addition of trifluoroacetic acid to norbornene is exceptionally fast at 0°C, the reaction requiring only 1–2 min for completion. The reaction product is 99.98% *exo*-norbornyl trifluoroacetate (49), 0.02% *endo*- (50) (20).

$$\tag{20}$$

Addition of trifluoroacetic acid-d_1 to norbornene again gave the product containing *exo*-3-d_1 (51) in excess over the Wagner–Meerwein-shifted product (52) (21). These additions are accompanied by the formation of

[23] P. E. Peterson, *J. Amer. Chem. Soc.*, **82**, 5834 (1960); P. E. Peterson, R. E. Kelley, Jr., R. Belloli, and K. A. Sipp, *ibid.*, **87**, 5169 (1965).

[24] J. E. Nordlander, R. R. Gruetzmacher, W. J. Kelly, and S. P. Jindal, *J. Amer. Chem. Soc.*, **96**, 181 (1974).

[25] P. E. Peterson and G. Allen, *J. Amer. Chem. Soc.*, **85**, 3608 (1963).

[26] P. E. Peterson, C. Casey, E. V. P. Tao, A. Agtarap, and G. Thompson, *J. Amer. Chem. Soc.*, **87**, 5163 (1965); P. E. Peterson, R. J. Bopp, D. M. Chevli, E. L. Curran, D. E. Dillard, and R. J. Kamat, *ibid.*, **89**, 5902 (1967).

much more hydride-shifted products than in the hydrogen chloride reaction
(17) (compare yields of **47** + **48** with **51** + **52**).

(21)

These results are again compatible with a carbonium ion process in
which a rapidly equilibrating pair, **45** ⇌ **46**, is captured prior to full equilibration.

The reaction of norbornene with deuteriotrifluoroacetic acid containing dissolved cesium chloride gives *exo*-norbornyl chloride as major product
(90%). The presence of the cesium chloride decreases the amount of
hydride-shifted material to as little as 5% in the reaction mixture. The
retained structure is favored, 61 : 34%. Thus the system behaves as though
the dissolved chlorides were trapping the equilibrating intermediate faster
than the less nucleophilic trifluoroacetate ion.

12.8. Addition of Perdeuterioacetic Acid to Norbornene

The greater nucleophilic properties of acetate ion and of acetic acid, as
compared to those of trifluoroacetate ion and of trifluoroacetic acid, lead to
the prediction that in the former medium a pair of rapidly equilibrating ions
should be captured much faster, making possible a more favorable synthesis
of the unrearranged isomer. It would also be anticipated that hydride-
shifted products would be greatly reduced. This is actually observed.[27]

The addition of acetic acid to norbornene proceeds relatively slowly at
100°C (22). The product is 99.98% *exo* (**53**) and 0.02% *endo* (**54**), a
stereoselectivity comparable to that realized in the acetolysis of the
arenesulfonates.[28]

[27]H. C. Brown and J. H. Kawakami, *J. Amer. Chem. Soc.*, **97**, 5521 (1975).
[28]H. L. Goering and C. B. Schewene, *J. Amer. Chem. Soc.*, **87**, 3516 (1965).

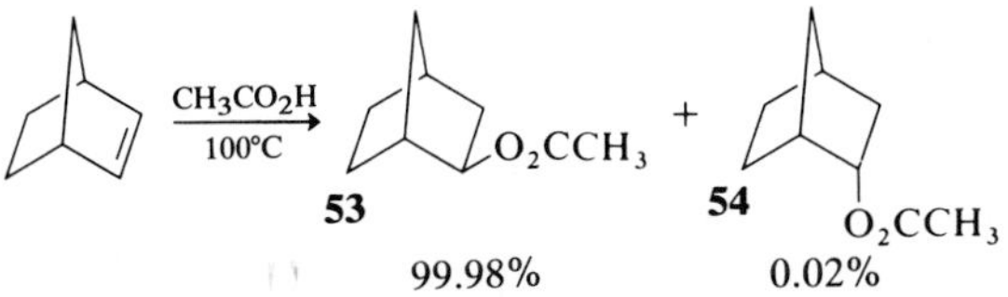

$$\text{(22)}$$

53 → 99.98% **54** 0.02%

Addition of perdeuterioacetic acid to norbornene provides 67% of the *exo*-3-*d* isomer (**55**) and 20% of *syn*-7-*d* (**56**), with a small amount of hydride-shifted products (23).

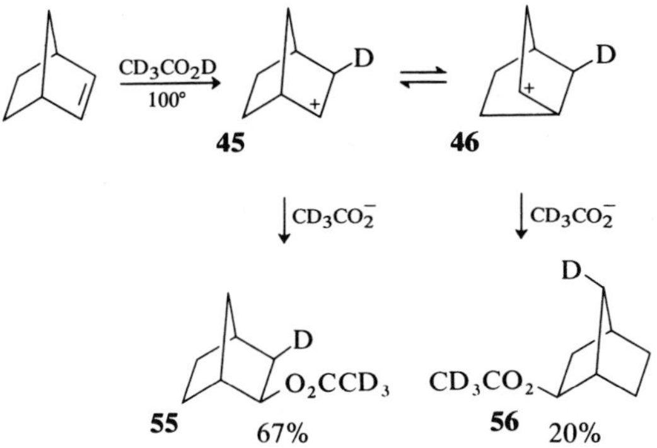

$$\text{(23)}$$

55 67% **56** 20%

These results create a difficulty for the suggestion that such additions may involve a considerable amount of concurrent molecular addition. To achieve the observed distribution would require that approximately 50% of the reaction goes through a carbonium ion process and 50% goes through the proposed cyclic addition. However, this proposal then requires that the cyclic addition process must proceed with an exceptionally high stereoselectivity, a stereoselectivity previously considered to be a diagnostic characteristic for the σ-bridged cation.[28]

The data clearly support the conclusion that in these addition reactions the rapidly equilibrating classical 2-norbornyl cations are being captured prior to full equilibration.

12.9. Resumé

It is clear that there are now a considerable number of experiments which point to the capture of the 2-norbornyl cation in an unsymmetrical form. Why have these results not received serious consideration?[10] The difficulty has been that the proposal that the 2-norbornyl cation is a symmetrical species has been accepted for so long that investigators have often refused to accept the actual implications even of their own experiments. Let us briefly summarize the results currently available.

1. Deamination of optically active *exo*-norbornylamine gives *exo*-norborneol with approximately 11–15% of residual activity.[7,9]

2. Addition of hydrogen chloride to 2,3-dideuterionorbornene gives a 56:42 excess of the isomer with retained structure (7).[18]

3. Reaction of 2,3-dideuterionorbornene with aqueous hydrobromic acid gives a 52:35 excess of the isomer with retained structure (9).[18]

4. The acid catalyzed reaction of 2,3-dideuterionorbornene with phenol produces the O-alkyl isomer with a 47:39 preference for the retained structure (10).[18]

5. The addition of deuterium chloride to norbornene at −78°C in methylene chloride solution produces a 57:41 distribution of isomers (17).[19] Again the retained structure is preferred.

6. Addition of deuteriotrifluoroacetic acid to norbornene at 0°C yields a 37:26 distribution favoring the retained structure (21).[22]

7. The reaction of norbornene with deuteriotrifluoroacetic acid containing dissolved cesium chloride gives *exo*-norbornyl chloride as major product[22] (90%) a decrease in hydride-shifted product and an increase in retained structure, 61:34%. The dissolved chloride behaves as a more effective nucleophile, trapping the rapidly equilibrating ions more effectively.

8. Addition of acetic acid to norbornene gives the *exo* isomer predominantly, 99.98% (22).[27] Addition of perdeuterioacetic acid gives largely the retained structure, 67:20 (23).

9. The acid-induced addition of perdeuterioacetic acid to norbornene likewise gives the retained structure preferentially,[19] 56:40.

10. Capture of the hydride shifted tagged 2-norbornyl cation, formed in the acetolysis of 2-(Δ^3-cyclopentenylethyl)-2-[14]C *p*-nitrobenzene-sulfonate, yield the Wagner–Meerwein acetates with a 38:25 preference for the retained structure (5).[13,14]

11. Finally, capture of the hydride-shifted cation from the acetolysis of *exo*-norbornyl-4-^{14}C tosylate reveals a 1.52:0.52 preference for the first member of the Wagner–Meerwein pair (6).[17]

12.10. Conclusion

The time is surely here when this question must be asked: Is it not solvolysis that is exceptional, rather than the numerous other carbonium ion reactions of the 2-norbornyl cation where the unsymmetrical (classical) structure has been successfully captured?

Comments

This chapter summarizes important observations. The reactions discussed definitely involve carbocationic intermediates, at least in part, since rearrangements are observed. Yet label scrambling is incomplete, suggesting to Brown that a classical norbornyl cation has been trapped before equilibration is complete. However, this is not the only possible interpretation, and I feel there is a basis for questioning the relevance of the results to the question of the structure of the 2-norbornyl cation. Let me point out possible difficulties.

Neither the transition state for deamination nor the transition state for the addition of protic acids to olefins possesses very much carbocation character. These are "carbonium ion reactions" to a lesser extent than solvolysis. Cationoid intermediates may be involved after the rate determining steps in both deamination and protic acid additions, but the evidence indicates that "tight ion pairs", rather than being "avoided" in such reactions, may even be *tighter* than those involved on solvolysis.[20,29] As Brown suggests, "the transition state for the protonation reaction involves much smaller development of positive charge than does that for solvolysis." Is it not more reasonable to expect that any intermediates formed after the transition state will also have less positive charge for protic acid addition than for solvolysis?

There is a fundamental test for a reaction intermediate: the same product spectrum should be obtained independent of the mode of generation of the species, providing that the reaction conditions are otherwise the same. However, different product mixtures or different extents of rearrangement are observed in deamination, protic acid addition, and solvolysis of norbornyl systems. If intermediates are present, they must differ significantly. Since the extent of label scrambling observed is greatest in solvolysis, it follows that the intermediates involved either have the most carbonium ion character and the least encumbrance by counter ion, nucleophile, and solvent, or have a larger lifetime. The results quoted in this chapter do not convince me that any species resembling a classical 2-norbornyl cation has been formed unambiguously and captured.

I believe it is an oversimplification to regard protic acid additions to norbornenes as proceeding [see (8), (17), and (23)] via stepwise addition of a proton to give a free carbocation intermediate, followed by nucleophilic attack. In some olefin additions, *termolecular* kinetics have been observed.[20]

---

[29]For a review and a discussion of mechanistic alternatives, see D. J. Raber, J. M. Harris, and P. v. R. Schleyer, *Ions and Ion Pairs in Organic Reactions*, Vol. II, M. Szwarc, Ed., Wiley, New York, 1974, Chapter 3, especially pp. 356–363.

Brown's investigations in this area have established neither the molecularity nor the mechanisms involved. For example, the operation of a termolecular process like **57** has not been excluded by kinetic studies as a possible explanation for the lack of complete label scrambling obtained upon addition of CF_3COOD to norbornene in the presence of CsCl.

57   **58**   **59**

Likewise, the *6-membered* molecular additions **58** and **59**, suggested by Fahey,[20] are not excluded by Brown's evidence.[21,30] It should be noted that **58** also represents the transition state for the reverse reaction, ester thermolysis.[31] The "Ad_E3 process" **60**,[20] which should not be hindred appreciably by the presence of a 7,7-dimethyl group, is another mechanistic possibility.

60

Transition states like **57–60** have sufficient polar character to give orientations following the Markovnikov rule. Compounds **57–60** may give products directly or may proceed further to give ion pairs capable of exhibiting rearrangement before final collapse to products occurs.

Deamination reactions have an "internal" pathway available which has no direct counterpart in tosylate or chloride solvolysis.[29,32–34] Consider the diazonium ion pair **61** showing one of the possible locations of the counter ion. Loss of N_2 occurs, leaving a new ion pair **62**. Even though the norbornyl cation may be bridged, the system **62** is unsymmetrical, because the anion is closer to C2 than to C1. If ion-pair collapse occurs before the counter ion has a chance to achieve a symmetrical location,[32] retention or partial retention

[30]W. Fliege and R. Huisgen, *Liebig's Ann.*, 2038 (1973).

[31]C. H. De Puy and R. W. King, *Chem. Rev.*, **60**, 431 (1960).

[32]J. T. Keating and P. S. Skell, *Carbonium Ions*, Vol. II, G. A. Olah and P. v. R. Schleyer, Eds., Wiley-Interscience, New York, 1970, Chapter 15.

[33]A. Streitwiesser, Jr., *J. Org. Chem.*, **22**, 861 (1957).

[34]E. L. Eliel, N. L. Allinger, S. J. Angyal, and G. A. Morrison, *Conformational Analysis*, Wiley-Interscience, New York, 1965, pp. 89f, 241–242.

of chirality in the product results (**63**). Mechanisms analogous to **61–63**, once termed S$_N$i, account, for example, for the large amount of equatorial product obtained on deamination of equatorial cyclohexylamines.[34] This reaction course is quite different from solvolysis results.

$$X^-N_2^+ \xrightarrow{-N_2} X^- \longrightarrow X$$

61   **62**   **63**

unsymmetrical
ion pair

    I wish to emphasize that the symmetry of the entire carbocation system, cation *and anion*, must be taken into account.[15,16] Collins' results,[17] for example, can be explained equally well by assuming that bridged norbornyl cations are present, but are biased towards one avenue of reaction because of the unsymmetrical location of the counter ion.

    In this chapter, Brown has provided strong evidence favoring the classical norbornyl cation. Other interpretations of his results are possible. I hope that further investigations will provide decisive answers.

13

Equilibrating Cations under Stable Ion Conditions

13.1. Introduction

In recent years the pioneering work of Olah, Saunders, Brouwer, and Hogeveen has made possible the direct spectroscopic observation of many cations.[1-3] It was natural to apply this new technique to the nonclassical ion problem. Indeed, Olah has concluded that "... the long standing controversy as to the nature of the norbornyl cation is unequivocally resolved in favor of the nonclassical carbonium ion."[4] Consequently, it appears appropriate to subject the studies and conclusions in this area to careful scrutiny to see if they support this conclusion.

One of the problems is that Olah has taken a number of intuitive positions which have not been tested objectively. First, he has applied results obtained under stable ion conditions to reinterpret conclusions previously reached on the basis of solvolytic studies without justifying this procedure. Second, he has utilized ^{13}C nmr shifts as a measure of charge delocalization, again without justifying this position. Third, he has utilized nonadditivity in ^{13}C nmr shifts as a basis for proposing nonclassical ion structures, again without testing this position. It appears desirable to examine these positions critically and objectively.[5] It is unfortunate that mere repetition of unsupported positions in a flood of publications can apparently convince readers that such positions must be valid.

[1]G. A. Olah, *Angew. Chem. Int. Ed. Engl.*, **12**, 173 (1973).
[2]M. Saunders, P. Vogel, E. L. Hagen, and J. Rosenfeld, *Accounts Chem. Res.*, **6**, 53 (1973).
[3]D. M. Brouwer and H. Hogeveen, *Progr. Phys. Org. Chem.*, **9**, 179 (1972).
[4]G. A. Olah, G. Liang, G. D. Mateescu, and J. L. Riemenschneider, *J. Amer. Chem. Soc.*, **95**, 8698 (1973).
[5]G. Kramer, *Adv. Phys. Org. Chem.*, **11**, 177 (1975).

13.2. Equilibrating Cations

As discussed earlier (Section 6.5), a major anomaly in carbonium ion chemistry was pointed out in 1965.[6] Systematic lowering of the potential barrier separating two symmetrical cations (1) would be expected to result in

$$\underset{R}{\overset{R}{>}}C-C< \;\; \rightleftharpoons \;\; >C-\underset{+}{C}< \tag{1}$$

three distinct classes of such cations: (A) essentially static classical cations, which can be formed and converted into products without significant equilibration; (B) equilibrating cations, which undergo rapid equilibration in the time interval between formation and conversion into products; and (C) bridged species where the potential barrier has disappeared so that resonance now occurs involving the two canonical structures. In 1965, practically all systems had been assigned to Classes A and C, with the intermediate Class B being essentially unpopulated. It was a puzzle why there should be this apparent discontinuity in the potential barrier separating such pairs of symmetrical cations (1).

This gradual transition is well illustrated in the methanolysis of the tagged 3-R-2,3-dimethyl-2-butyl-3,5-dinitrobenzoates (1) (2).[7]

For R = methyl, no scrambling is observed in the methyl ether product (4). Evidently the cation 2 is captured before it can undergo a Wagner–Meerwein transformation into 3. Thus the 2,3,3-trimethyl-2-butyl system under these conditions is properly placed in Class A.

For R = phenyl, the ether product reveals 45% scrambling (77.5% 4, 22.5% 5). Evidently the rate of capture of the cations, 2⇌3, is somewhat faster than the rate of equilibration. Thus the 3-phenyl-2,3-dimethyl-2-butyl system under these conditions is properly classified as B.

[6]H. C. Brown, K. J. Morgan, and F. J. Chloupek, *J. Amer. Chem. Soc.*, **87**, 2137 (1965).
[7]H. C. Brown and C. J. Kim, *J. Amer. Chem. Soc.*, **90**, 2082 (1968).

For R = *p*-anisyl, the product (50% **4**, 50% **5**) reveals complete scrambling. The cation is either aryl-bridged (Class C), or involves a rate of equilibration that is rapid compared to the rate of capture (Class B). (In other carbonium ion reactions of the *p*-anisyl derivative, the intermediates, **2**⇌**3**, were apparently captured prior to full equilibration.[7])

The situation has now been altered. Study of carbonium ions (1) under stable ion conditions has established that rapid equilibration is the norm. Even cations such as 2,3,3-trimethyl-2-butyl, which can be captured in solvolytic processes without equilibration, as just discussed, undergo very rapid equilibration under stable ion conditions. Such equilibration often cannot be frozen out even at temperatures as low as −150°C.

Representative examples are summarized in Table 13.1.

TABLE 13.1. *Representative Equilibrating Classical Cations under Stable Ion Conditions*

Cation	Structure	Assignment	Reference
2-Butyl	C–C–C–C ⇌ C–C–C–C	Equil. class	Brouwer[a] Olah[b] Saunders[c]
2,3-Dimethyl-2-butyl	C–C–C ⇌ C–C–C	Equil. class	Brouwer[a] Olah[b] Saunders[c]
1,2-Dimethyl-cyclopentyl		Equil. class	Olah[d]
Bicyclo[3.3.0]-octyl		Equil. class	Olah[d]
Bicyclo[4.4.0]-decyl		Equil. class	Olah[d]
2,3,3-Trimethyl-2-butyl	C–C–C–C ⇌ C–C–C–C	Equil. class	Brouwer[a] Olah[b] Saunders[c]

TABLE 13.1 (*continued*)

Cation	Structure	Assignment	Reference
Cyclopentyl	(cyclopentyl cation equilibria) ... etc.	Equil. class	Brouwer[a] Olah[b]
1-Methylcyclo-butyl	(1-methylcyclobutyl cation ⇌ structure)	Equil. class	Saunders[e]
2,4-Dimethyl-2-pentyl	(structure with H, C, C, C, C ⇌ C, C, C, C)	Equil. class	Brouwer[a] Saunders[c]
2,3-Dimethyl-2-norbornyl	(norbornyl cation with H ⇌ norbornyl cation with H)	Equil. class	Olah[f] Sorensen[g]
1,2-Dimethoxy-2-norbornyl	(norbornyl cation OCH_3 OCH_3 ⇌ OCH_3 OCH_3)	Equil. class	Nickon[h]
1,2-Di-*p*-anisyl-2-norbornyl	(norbornyl cation *p*-An *p*-An ⇌ *p*-An *p*-An)	Equil. class	Schleyer[i]
1,2-Diphenyl-2-norbornyl	(norbornyl cation Ph Ph ⇌ Ph Ph)	Equil. class	Olah[j]
1,2-Dimethyl-2-norbornyl	(norbornyl cation ⇌ norbornyl cation)	Equil. class	Olah[k]

TABLE 13.1 (*continued*)

Cation	Structure	Assignment	Reference
2-Norbornyl	[structure] ⇌ or ⟷ [structure]	Nonclassical	Olah[l]
Cyclopropyl-carbinyl	[structure] ⇌ or ⟷ [structure]	Nonclassical Equil. class	Olah[m] Kelly[n]

[a] Reference 3. [b] Reference 1. [c] Reference 2. [d] G. A. Olah, G. Liang, and P. W. Westermann, *J. Org. Chem.*, **39**, 367 (1974). [e] M. Saunders and J. Rosenfeld, *J. Amer. Chem. Soc.*, **92**, 2548 (1970). [f] G. A. Olah and G. Liang, *J. Amer. Chem. Soc.*, **96**, 189 (1974). [g] A. J. Jones, E. Huang, R. Haseline, and T. S. Sorensen, *J. Amer. Chem. Soc.*, **97**, 1133 (1975). [h] A. Nickon and Y. Lin, *J. Amer. Chem. Soc.*, **91**, 6861 (1969). [i] P. v. R. Schleyer, D. C. Kleinfelter, and H. G. Richey, Jr., *J. Amer. Chem. Soc.*, **85**, 479 (1963). [j] G. A. Olah and G. Liang, *J. Amer. Chem. Soc.*, **96**, 195 (1974). [k] Reference 42. [l] References 4, 13, 27, 44. [m] References 35, 36. [n] Reference 37.

Consequently of all of these carbonium ions, for many of which nonclassical ionic structures were previously considered, only cyclopropyl-carbinyl and 2-norbornyl are still assigned σ-bridged structures on the basis of their spectroscopic characteristics under stable ion conditions.

It should be pointed out that nonclassical structures have been proposed for additional systems.[8-10] Some of these systems will be considered in Chapter 14. These systems have not yet been subjected to careful scrutiny. On the other hand, we do have an immense amount of data for the cyclopropylcarbinyl and 2-norbornyl cations. If we cannot now analyze these data and arrive at an understanding of these systems satisfactory to all, with the immense amount of data now available, there appears to be little point in extending the discussion to additional systems for which the availability of comparable data is minute.

13.3. Applicability of Results to Solvolysis

We are faced with two questions. Do the studies of cations under stable ion conditions really support σ-bridged structures for the 2-norbornyl and cyclopropylcarbinyl cations as proposed?[1] Are results for carbonium ions realized under stable ion conditions directly applicable to the structures and

[8] G. Seybold, P. Vogel, M. Saunders, and K. B. Wiberg, *J. Amer. Chem. Soc.*, **95**, 2045 (1973).
[9] D. Lenoir, D. J. Raber, and P. v. R. Schleyer, *J. Amer. Chem. Soc.*, **96**, 2149 (1974).
[10] H. Hogeveen and P. W. Kwant, *Accounts Chem. Res.*, **8**, 413 (1975).

behavior of these cations under solvolytic conditions? Let us consider the latter question first.

The question is one of long standing and was raised by Winstein in the early days of the observation of cations under stable ion conditions.[11] "While we regard direct observations of such intermediates [cations] as very desirable, we disagree with Deno,[12] who states that such direct observation should render 'obsolete' the past solvolytic work. Solvation and reaction possibilities are so different in the typical solvolyzing solvents than in the inert non-solvolyzing ones that both types of investigation are essential."

Olah's position on this point appears to have varied. For example, in 1970 he pointed out that his conclusions as to the symmetrically bridged structure of the 2-norbornyl cation under stable ion conditions might not apply to the cation under solvolytic conditions.[13] Yet in 1973 he exhibited no hesitation in making this extrapolation,[4] as was pointed out earlier.

The need for such caution is revealed by the contrasting behavior of the 2,3,3-trimethyl-2-butyl cation under solvolytic[7] and stable ion conditions.[14] Under such conditions the cation is rapidly equilibrating, revealing in nmr only one type of methyl group. The equilibration is not frozen out even at temperatures as low as $-180°C$, with an energy of activation for the rearrangement of the methyl group ≤ 3 kcal mol^{-1}.[14]

On the other hand, methanolysis of the corresponding tagged p-nitrobenzoate (2)[7] yields the methyl ether 5 without detectable scrambling of the tag. If we take the limit of detection of scrambling to be $\sim 1\%$, then in methanol the energy of activation for the capture of the cation must be ~ 3.5 kcal mol^{-1} less than the energy of activation for the scrambling of the methyl groups. If we assume that this quantity is the same in methanol as in superacids, we are left with the deduction that the reaction of the cation with solvent involves zero energy of activation or less. Yet such tertiary cations reveal considerable discrimination between different nucleophiles.[15] Clearly, the assumption made is invalid (Section 6.5).

A study of this question in the 2-aryl-2-norbornenyl derivatives has recently been made.[16] In solvolysis, examination of the effect of increasing electron demand reveals no significant change in the *exo*:*endo* rate ratio with p-CH$_3$O, p-H, and p-CF$_3$ substituents.[17] Only with the more demand-

[11]R. Howe, E. C. Friedrich, and S. Winstein, *J. Amer. Chem. Soc.*, **87**, 379 (1965).

[12]N. C. Deno, *Prog. Phys. Org. Chem.*, **2**, 129 (1964).

[13]G. A. Olah, A. M. White, J. R. De Member, A. Commeyras, and C. Y. Lui, *J. Amer. Chem. Soc.*, **92**, 4627 (1970).

[14]G. A. Olah and J. Lukas, *J. Amer. Chem. Soc.*, **89**, 4739 (1967).

[15]R. A. Sneen, J. V. Carter, and P. S. Kay, *J. Amer. Chem. Soc.*, **88**, 2594 (1966).

[16]D. G. Farnum and R. E. Botto, *Tetrahedron Lett.*, 4013 (1975).

[17]H. C. Brown and E. N. Peters, *J. Amer. Chem. Soc.*, **97**, 7442 (1975).

ing 3,5-$(CF_3)_2$ substituent does the *exo*:*endo* rate ratio exhibit an increase attributable to π-participation.

On the other hand, in FSO_3H the nmr spectra reveal changes indicating a change from no π-participation with *p*-CH_3O to participation with *p*-H, *p*-CF_3, and 3,5-$(CF_3)_2$. Thus the onset of π-participation is observed in this series at a much earlier point than it is in solvolysis. The authors agree to the need for caution in extrapolating data from superacid media to solvolytic media.

Finally, it is instructive to compare results and conclusions for the solvolysis of tagged 2,3-dimethyl-3-aryl-2-butyl chlorides (**6**) in methanol (3)[18] with Olah's successive conclusions as to the structure of the 3-aryl-2,3-dimethyl-2-butyl cation under stable ion conditions.

The results on the observed scrambling are summarized in Table 13.2.

The results for *p*-methoxy could be attributed to the formation of a bridged intermediate. However, in the other three cases complete equilibration of the tag did not occur. Indeed, the less activating the substituent, the less scrambling observed in the product. Consequently, the results are most simply accounted for in terms of a rapid equilibration (**7** $\rightleftharpoons$ **8**), whose rate varies in the order: *p*-$CH_3O > p$-$CH_3 > p$-H $> p$-Cl. The rate of equilibration for *p*-H and *p*-Cl is less than the rate of capture of the equilibrating cation by solvent (methanol).

The 3-aryl-2,3-dimethyl-2-butyl cations have also been examined under stable ion conditions. What conclusions have been drawn as to the structures of these cations?

[18]Unpublished data with C. J. Kim.

TABLE 13.2. *Products from the Methanolysis of the Substituted 2,3-Dimethyl-3-aryl-2-butyl Chlorides at 50°C*

Substituent Z	Product				Scrambling, %	
	Methyl ether		Olefin			
	Yield, %	Rearr., %	Yield, %	Rearr., %	Ether	Olefin
p-Methoxy[a]	72	50[a]	28	62[a]	100	100
p-Methyl	34	49	66	48	97	95
p-Hydrogen	25	28	75	27	49	47
p-Chloro	17	17	83	17	19	19

[a] Accuracy reduced because initial chloride was 29% scrambled in synthesis.

1. The 2,3-dimethyl-3-phenyl-2-butyl cation is either open or bridged.[19]
2. The 2,3-dimethyl-3-phenyl-2-butyl cation undergoes rapid equilibration with no evidence for a bridged ion.[20]
3. The 2,3-dimethyl-3-phenyl-2-butyl cation exists as the phenonium ion—the first example of a stable long-lived phenonium ion without any ring substituents.[21]
4. On the other hand, the p-CF$_3$ derivative exists under the same conditions as a rapidly equilibrating classical cation.[21]
5. The CF$_3$ derivative exists as the bridged phenonium ion.[22]

What conclusions can we draw? First, it is evident that the behavior of cations under stable ion conditions does not necessarily correlate with their behavior under solvolytic conditions. Second, until there is better understanding of the observed spectra of cations under stable ion conditions, caution should be exercised in accepting conclusions as to the structure of such cations based on such observations alone.

13.4. ^{13}C NMR Shifts as a Measure of Charge Delocalizations

The study of solvolysis has provided a huge mass of information as to the influence of structure on the stability of transition states.[23] Since such transition states are closely related to the corresponding cations,[24] the

[19] G. A. Olah and C. U. Pittman, Jr., *J. Amer. Chem. Soc.*, **87**, 3509 (1965).

[20] G. A. Olah, C. U. Pittman, Jr., E. Namanworth, and M. B. Comisarow, *J. Amer. Chem. Soc.*, **88**, 5571 (1966).

[21] G. A. Olah, M. B. Comisarow, and C. J. Kim, *J. Amer. Chem. Soc.*, **91**, 1458 (1969).

[22] G. A. Olah and R. D. Porter, *J. Amer. Chem. Soc.*, **93**, 6877 (1971).

[23] A. Streitwieser, Jr., *Solvolytic Displacement Reactions*, McGraw-Hill, New York, 1962.

[24] G. S. Hammond, *J. Amer. Chem. Soc.*, **77**, 334 (1955).

results have been utilized to establish the effect of structure upon the stabilities of cations. A broad consistent understanding has been realized.

For example, the rate of solvolysis of 2-methyl-*exo*-norbornyl *p*-nitrobenzoate (**11**) is faster than that of 1-methylcyclopentyl *p*-

	11 CH$_3$	**12** CH$_3$	(4)
RR (25°C):	4.74	1.00	

nitrobenzoate (**12**) by a relatively small factor (4.74).[25] On the assumption that there are no major differences in ground-state energies, or other unusual structural effects, one would conclude that the stabilization of the two cationic centers by the organic moieties must be comparable in these two systems. This conclusion is consistent with the conclusion that the 2-methyl-2-norbornyl cation must be classical, as confirmed by the constant effect of a 1-methyl substituent in the *exo* and *endo* isomers of 1,2-dimethyl-2-norbornyl (Section 9.5), and by the successful capture of the optically active cation.[26]

More recently, Olah has proposed to relate the ^{13}C chemical shift of the carbonium ion carbon atom to the amount of positive charge at the central carbon atom in a cation by use of the "well-established relationship of ^{13}C chemical shifts to electron density."[27]

For example, is it reasonable to ascribe a partial nonclassical (σ-bridged) structure to the 2-methyl-2-norbornyl cation (**15**) on the basis of a difference in the ^{13}C shifts of the 2-methyl-2-norbornyl (**13**) and 1-methylcyclopentyl cations (**14**)?[27]

	13 CH$_3$	**14** CH$_3$	**15** CH$_3$
[13]C shift of C$^+$	269.9	336	

As it happens, a careful check of the literature fails to reveal such a "well-established relationship of ^{13}C chemical shifts to electron density."[27a] Indeed, the validity and scope of such a relationship in carbonium ion chemistry would appear highly questionable. For example, the basis of

[25]S. Ikegami, D. L. Vander Jagt, and H. C. Brown, *J. Amer. Chem. Soc.*, **90**, 7124 (1968).

[26]H. L. Goering and J. V. Clevenger, *J. Amer. Chem. Soc.*, **94**, 1010 (1972).

[27]G. A. Olah and A. M. White, *J. Amer. Chem. Soc.*, **91**, 5801 (1969).

[27a]For a recent discussion of the complexities of ^{13}C shifts, see D. G. Farnum, *Adv. Phys. Org. Chem.*, **11**, 123 (1975).

this relationship apparently rests on data by Lauterbur, which suggest that variations in π-electron densities are primarily responsible for ^{13}C chemical shifts.[28] In fact, the ^{13}C chemical shifts of charged aromatic derivatives have been shown to be related to the local π-electron density.[29] However, it has not yet been demonstrated that this relationship for π-electron systems holds true for alkyl carbonium ions. Indeed, on an *a priori* basis one would not expect any relationship between ^{13}C chemical shifts and electron density to be valid in compounds which are very different from one another since ^{13}C chemical shifts have been reported to depend upon changes in the paramagnetic screening term centered on the atom of interest.[30] It has been suggested that the paramagnetic screening term depends on several important factors, such as: (a) the excited states of the atom; (b) electron polarization; (c) deviations from classical perfect-pairing bond structures; (d) steric polarization of electrons along an H–C bond; and (e) steric crowding.[30]

Examination of the ^{13}C chemical shifts of *tert*-cumyl cations reveals the existence of a linear relationship with $\sigma+$ constants.[31] However, many reactions involving changes in substituents in the *meta* and *para* positions of the aromatic ring are correlated by the Hammett relationship, but the same kind of quantitative treatment cannot be carried over to other systems where structural variations are made at or near the reaction center. Consequently, it is desirable to examine other systems critically[5] to test whether it is valid to use ^{13}C shifts as Olah and his coworkers have done.

Certain data suggest the hazards involved in this procedure. The isopropyl cation is far less stable than the *tert*-butyl cation and must possess a much higher concentration of charge at the cationic carbon. The *tert*-butyl cation exhibits a ^{13}C chemical shift of 329.2 ppm; consequently, the isopropyl cation would be expected to exhibit an even greater ^{13}C shift. However, the observed value is not larger, but smaller, 318.8 ppm.[5] Thus, Olah deduces that "the well-established relationship of ^{13}C shifts to electron density leads to the conclusion that the central carbon atom in the *tert*-butyl cation is slightly more positive than that in the isopropyl cation."[27]

This apparent anomaly might be a consequence of the tendency for such highly active secondary species to exist as tight ion pairs. If so, the anomaly might vanish with more stable cationic species. Both thermodynamic and kinetic data reveal the triphenylmethyl cation to be more stable than the diphenylmethyl cation.[32] However, the observed chemical shifts are 211.9

[28]P. G. Lauterbur, *J. Amer. Chem. Soc.*, **83**, 1838, 1846 (1961).

[29]H. Spiesecke and W. G. Schneider, *Tetrahedron Lett.*, 468 (1961).

[30]B. V. Cheney and D. M. Grant, *J. Amer. Chem. Soc.*, **89**, 5319 (1967).

[31]G. A. Olah, C. L. Jeuell, and A. M. White, *J. Amer. Chem. Soc.*, **91**, 3961 (1969).

[32]N. C. Deno and A. Schriesheim, *J. Amer. Chem. Soc.*, **77**, 3051 (1955).

and 199.4 ppm, respectively.[27] Thus, even these relatively stable cations do not obey the proposed relationship.

Olah and his coworkers examined the *tert*-butyl (16), phenyldimethyl-(17), and cyclopropyldimethylcarbonium (18) ions and reported ^{13}C chemical shifts of 329.2, 254.9, and 280.6 ppm, respectively.[27] They proposed from these results that the phenyl group must be more electron releasing than a cyclopropyl group.

	16	17	18
[13]C shift for C[+]	329.2	254.9	280.6
RR (25°C):	1.00	969	503,000

This conclusion runs counter to a large number of experimental data, including Olah's own pmr results.[33] Indeed, an examination of the rates of solvolysis of the corresponding *p*-nitrobenzoates reveal reactivities that indicate cyclopropyl to be a far better stabilizing group than phenyl (5).[34]

Clearly, it is desirable to proceed with caution in basing conclusions as to the electron densities of cationic centers on ^{13}C shifts alone until the factors which influence such shifts are better understood.

13.5. Nonadditivity of [13]C Shifts as a Basis for Nonclassical Structures

Olah has utilized discrepancies between the observed ^{13}C shifts and the values calculated for a set of rapidly equilibrating classical cations to argue for σ-bridged structures for the cyclopropylcarbinyl[35,36] and 2-norbornyl cations.[13] Consequently, it is appropriate to examine this argument critically. Fortunately, it proved especially convenient to test the proposal for the cyclopropylcarbinyl cation, and this study[37] will be presented.

Olah and his coworkers concluded that the cyclopropylcarbinyl cation must exist as a nonclassical σ-bridged species on the basis of a discrepancy between the observed ^{13}C shift and the value calculated for a set of rapidly equilibrating classical cations.[35,36] This is a surprising conclusion in view of

[33]C. U. Pittman, Jr., and G. A. Olah, *J. Amer. Chem. Soc.*, **87**, 5123 (1965).
[34]H. C. Brown and E. N. Peters, *J. Amer. Chem. Soc.*, **95**, 2400 (1973).
[35]G. A. Olah, D. P. Kelly, C. L. Jeuell, and R. D. Porter, *J. Amer. Chem. Soc.*, **92**, 2544 (1970).
[36]G. A. Olah, C. L. Jeuell, D. P. Kelly, and R. D. Porter, *J. Amer. Chem. Soc.*, **94**, 146 (1972).
[37]D. P. Kelly and H. C. Brown, *J. Amer. Chem. Soc.*, **97**, 3897 (1975).

the huge amount of evidence that the cyclopropyl group best stabilizes the cationic center in the bisected form (**19**), whereas a σ-bridged structure would require the less favorable parallel conformation (**20**).

There are major uncertainties in the argument. First, the ^{13}C shift for the carbonium carbon of the primary cation in cyclopropylcarbinyl is estimated from a linear extrapolation of the values for the nonequilibrating secondary and tertiary cyclopropylcarbinyl cations. It has not been established that such a linear extrapolation is valid. Second, as pointed out in the last section, the factors influencing ^{13}C shifts appear to be complex, and are not simply related to charge density or structure. Third, it is not possible to test the conclusion by calculating and demonstrating agreement of the observed ^{13}C shift with the proposed σ-bridged structure. No basis now exists for such calculations.

Fortunately, in the case of cyclopropylcarbinyl, another independent nmr criterion was suggested for the formation of a σ-bridge. Such a σ-bridge would cause the carbonium carbon to move toward the cyclopropyl ring. There is considerable evidence that such decrease in the dihedral angle and increase in strain will be reflected in major increases in the ^{13}C–H coupling constants, as shown by **21**, **22**, and bicyclobutane (**23**) itself.[38]

The same effect is shown in the bicyclobutane derivatives, **24–26**, where the increase in strain increases the apical methine coupling from 200 Hz in **24** to 212 Hz in **26**.

It was on this basis that it was predicted that the coupling constant for the apical methine hydrogen would increase from the normal values anticipated for cyclopropyldimethylcarbinyl (**27**) and cyclopropylmethylcarbinyl (**28**) to a much higher value, in the neighborhood of 200 Hz, for the

[38]J. B. Stothers, *Carbon-13 NMR Spectroscopy*, Academic Press, New York, 1972.

σ-bridged primary ion (**29**). However experiment does not bear out that prediction.[37] The value reveals a small decrease rather than the increase expected.

$\begin{matrix} H \\ 187 \end{matrix}$ **27** $\begin{matrix} H \\ 190 \end{matrix}$ **28** $\begin{matrix} H \\ 180 \end{matrix}$ **29**

These results have important implications. Two major nmr criteria have been proposed by Olah to distinguish the formation of σ-bridged nonclassical ions in superacids. The first is the agreement between calculated and observed values for ^{13}C shifts. A failure to obtain agreement has been used as a basis to argue for the formation of a σ-bridged species.[13,36] The second has been an increase in the ^{13}C–H coupling constant postulated to accompany an increase in strain with the formation of the σ-bridge.[35] Failure to observe a substantially different coupling constant has been used to argue for a classical cationic intermediate.[39]

In the case of the cyclopropylcarbinyl cation, these criteria now lead to conflicting interpretations. The earlier chemical shift data led to a preference for the σ-bridged structure, in spite of the mass of data favoring the bisected conformation in this system.[36] The present coupling-constant data support a structure without such a σ-bridge. Clearly it is necessary to proceed with caution before using either nmr criterion as a basis for a decision.

Disagreement between the calculated value for the ^{13}C shift of C1 and C2 in equilibrating classical ions, 185 ppm, and the observed value, 125.3 ppm, has been used to argue for the nonclassical structure of 2-norbornyl[13] (Section 6.2: **3**, **4**, **5**, **6**, **7**, **8**, or **9**). However, this procedure suffers from the difficulties previously discussed. The ^{13}C shift for the static 2-propyl cation is used as an estimate for the ^{13}C shift to be expected for the static classical 2-norbornyl cation. This is an enormous extrapolation. Kramer has pointed out that the value for 2-propyl cannot be utilized in this way to calculate the observed shift in the equilibrating classical 2-butyl cations[5] (Table 13.1). If this relatively minor extrapolation causes difficulty, what confidence can be placed on the much greater extrapolation involved in using the 2-propyl cation as a model for 2-norbornyl?

13.6. PMR Spectra

The pmr spectra of the 2-norbornyl cation reveal fascinating features, and attempts have been made to utilize those spectra to support the

[39]G. A. Olah, G. Liang, K. A. Babiak, and R. K. Murray, *J. Amer. Chem. Soc.*, **96**, 6794 (1974).

σ-bridged structure. The ion, prepared from *exo*-norbornyl fluoride in SbF_5-SO_2[40] or from *exo*-norbornyl bromide in $GaBr_3$-SO_2,[41] exhibits at room temperature in its pmr spectrum a single broad band at δ 3.10, attributed to scrambling of all of the hydrogen atoms through rapid Wagner–Meerwein 1:2 rearrangement and both 3:2 and 6:2 hydride shifts.

At $-70°C$, three individual peaks appear with relative areas of 4:1:6.[40,41] No additional change occurred down to $-120°C$.[13]

By utilizing a mixed SbF_5-SO_2ClF-SO_2F_2 solvent, it proved possible to observe the 100 MHz spectrum down to $-154°C$.[13] The low-field peak of area 4 separates into two peaks, each of area 2, at δ 3.05 and δ 6.59. The high field resonance attributed to the six methylene protons (on C3, C7 and C5) broadens, developing a shoulder at δ 1.70. The peak at δ 2.82 attributed to the bridgehead proton at C4 remains unchanged.

These changes in the spectrum are attributed to the freezing out of the 3:2 hydride shift (E_{act} 10.8 kcal mol^{-1}) below $-30°C$, with the 6:2 hydride shift and the Wagner–Meerwein 1:2 rearrangement continuing to be rapid on the nmr time scale even at $-120°C$. Below that temperature, the 6:2 hydride shift is frozen out (E_{act} 5.9 kcal mol^{-1}), but the Wagner–Meerwein rearrangement is not.

Is a Wagner–Meerwein rearrangement that is fast on the nmr time scale at $-154°C$ compatible with a classical structure for the cation? Or does it require a σ-bridged structure? The answer is provided by Table 13.1. The Wagner–Meerwein rearrangement of the 2,3,3-trimethyl-2-butyl cation is not frozen out at $-180°C$.[14] Yet is is assigned a classical structure.[14] Similarly, the Wagner–Meerwein rearrangement of the 1,2-dimethyl-2-norbornyl cation is not frozen out at $-140°C$.[42] Consequently, the fast rate of the Wagner–Meerwein rearrangement, in itself, is not a satisfactory argument for a σ-bridged structure.

It has been argued that the unusually slow 3:2 hydride shift supports the σ-bridged structure.[41] However, it has been pointed out that the slow 3,2-hydride shift may also arise from poor alignment of the *p*-orbital of the secondary ion and the sp^3 orbital of the *exo*-C–H bond involved in the hydride transfer.[5] This explanation has been utilized to account for the slow interconversion of the tertiary 1- into the secondary 2-adamantyl cation.[43]

We can only conclude that the pmr spectra fail to provide convincing evidence for the σ-bridged structure of the 2-norbornyl cation.[5]

[40]M. Saunders, P. v. R. Schleyer, and G. A. Olah, *J. Amer. Chem. Soc.*, **86**, 5680 (1964).

[41]F. R. Jensen and B. H. Beck, *Tetrahedron Lett.*, 4287 (1966).

[42]G. A. Olah, J. R. De Member, C. Y. Lui, and R. D. Porter, *J. Amer. Chem. Soc.*, **93**, 1442 (1971).

[43]D. M. Brouwer and H. Hogeveen, *Rev. Trav. Chim.*, **89**, 211 (1970).

13.7. ESCA Spectra

The C-1s X-ray electron spectroscopic examination of the 2-norbornyl cation offers perhaps the best hope of establishing the structure of the 2-norbornyl cation under stable ion conditions. (Whether such results will be applicable to the species formed under solvolytic conditions is another question, as discussed in Section 13.3.) Olah has reported two ESCA spectra[4,44] for the 2-norbornyl cation. It was this study that led him to conclude that "... the long standing controversy as to the nature of the norbornyl cation is unequivocally resolved in favor of the non-classical carbonium ion."[4] However, the experimental evidence offered for the σ-bridged structure by the ESCA spectra has been questioned by Kramer.[5]

In principle the ESCA spectra should be capable of providing an unambiguous answer to the structure of the 2-norbornyl cation under stable ion conditions. The rapidly equilibrating classical cation should exhibit two peaks with an area ratio of 6:1, whereas the nonclassical structure should exhibit two peaks with an area ratio of 5:2.

It was reported that the *tert*-butyl and the cyclopentyl cation give ESCA spectra in which the peak representing the cationic center is shifted about 4 eV to the high binding energy side of a single band representing the other carbon atoms, the shift being somewhat larger in the secondary than in the tertiary system.

The 2-norbornyl spectrum reveals a much smaller shift, estimated as 1.7 eV. This, together with the claim that the area ratio was indeed 5:2 appeared to provide persuasive evidence in favor of the nonclassical ion.

This position is beclouded by three developments. First, Dewar has reported that MINDO/3 examination of the 2-norbornyl cation leads him to the conclusion not only that it is classical, but that the classical ion would lead to the observed ESCA spectra.[45] Second, Kramer has reported that a detailed analysis of the original published curve[44] reveals an area considerably closer to 6:1 than 5:2.[5] The later spectrum[4] appears to contain a new high binding energy band on the shoulder of what was supposedly the cationic band. If the area of this band is deleted, the remaining bands again yield a 6:1 area ratio.[5] Third, there appear to be major uncertainties in the experimental procedure. Thus, in discussions with Olah and Mateescu, they have indicated that the experimental technique is a difficult one and that

[44]G. A. Olah, G. D. Mateescu, and J. L. Riemenschneider, *J. Amer. Chem. Soc.*, **94**, 2529 (1972).

[45]M. J. S. Dewar, R. C. Haddon, A. Komornicki, and H. Rzepa, *J. Amer. Chem. Soc.*, **99**, 377 (1977).

only one out of many spectra is "acceptable." Regrettably, these difficulties were not reported.[4,44,46]

Indeed, other workers have encountered major difficulties in attempting to obtain ESCA spectra of carbonium ions. For example, Professor Martin Saunders of Yale University has commented to the author: "I agree with you that the published ESCA data are not conclusive. In our own ESCA studies we have observed that the peaks vary considerably from experiment to experiment. I believe that the variability arises from surface effects since the ESCA experiments only looks at the first 30 Å of material. It is extremely difficult to prevent small amounts of contamination from entering an ESCA instrument during the sample preparation or afterward. These impurities could very easily destroy carbonium ions. It is regrettable that no one has yet repeated any of Olah's ESCA experiments in this area."

Finally, in addition to these technical difficulties, there is a major uncertainty in the procedure which has not yet received attention. It has recently been discovered that such solutions of carbonium ions can involve a number of equilibria.[47] Consequently, there is no certainty as to the precise solid phase which precipitates on the probe and is then subjected to ESCA examination.

The ESCA technique, in principle, should be capable of providing an unambiguous answer to the structure of the 2-norbornyl cation under stable ion conditions. Regrettably, the results available at the present time do not inspire confidence that such a solution is at hand.

13.8. Cation Stabilities

It appears appropriate to use rates of solvolysis of secondary tosylates in trifluoroacetic acid, a solvent of exceptionally low nucleophilicity,[48] to estimate the relative stability of cations. The available data[48,49] indicate

[46]G. A. Olah, *Accounts Chem. Res.*, **9**, 41 (1976). In this review Olah does describe some of the experimental difficulties, but does not reveal why these difficulties were not reported in the original papers. It is noteworthy that other workers have encountered major difficulties in obtaining such ESCA spectra, such as H. Hogeveen and P. W. Kwant, *Accounts Chem. Res.*, **8**, 413 (1975). It is disturbing that five years after the first ESCA spectra were reported,[44] no other workers in other laboratories have been successful in realizing such spectra. Finally, a critical comparison of the Olah review with the present discussion will clarify for the reader some of the difficulties that have been experienced in resolving the fascinating question of the structure of the 2-norbornyl and related cations.

[47]G. A. Olah and Y. K. Mo, *J. Amer. Chem. Soc.*, **96**, 3560 (1974).

[48]P. E. Peterson and F. J. Slama, *J. Amer. Chem. Soc.*, **90**, 6516 (1968).

[49]J. E. Nordlander, R. R. Gruetzmacher, W. J. Kelly, and S. P. Jindal, *J. Amer. Chem. Soc.*, **96**, 181 (1974).

increasing order of stability: 2-propyl$^+$ < cyclopentyl$^+$ < 2-norbornyl$^+$. We can extend this order to the tertiary derivatives, where solvent participation is a less significant factor: 2-norbornyl$^+$ < *tert*-butyl$^+$ < 1-methylcyclopentyl$^+$ < 2-methyl-2-norbornyl$^+$.[50]

According to these data, *exo*-norbornyl is a relatively reactive secondary, but considerably less reactive than representative tertiary derivatives, such as *tert*-butyl itself. In accordance with the usual procedure of proceeding from solvolytic data to estimates of carbonium ion stability (Section 13.3), these results point to a considerably greater stability of the *tert*-butyl cation than of the secondary 2-norbornyl cation.

In recent years a number of workers have reported data for the gas phase[51,52] and for superacid media which indicate the 2-norbornyl cation to be either considerably more stable than *tert*-butyl,[51,52] or of comparable stability.[53,54] They have attributed this change in the usual stabilities of secondary and tertiary cations to the stabilization provided by σ-bridging in the 2-norbornyl cation.

Fortunately, Arnett and his coworkers have developed a calorimetric procedure for measuring the heats of ionization of alkyl chlorides in methylene chloride solution (6):[55]

$$\text{RCl} + \text{SbF}_5 \xrightarrow[-25°\text{C}]{\text{CH}_2\text{Cl}_2} \text{R}^+ + \text{SbF}_5\text{Cl}^- \tag{6}$$

Their preliminary results indicate that ΔH for the ionization of *exo*-norbornyl chloride is -11.0 kcal mol^{-1}, as compared to a considerably more exothermic reaction, -16.3 kcal mol^{-1} for *tert*-butyl chloride.[55] These results are in agreement with the conclusions based on the solvolytic data and raise questions as to the validity of the alternative approaches.[51-54]

13.9. Conclusion

The development of methods for preparing and observing carbonium ions under stable ion conditions is a promising advance. It would be highly interesting to know the precise structure of the 2-norbornyl cation under these conditions. Unfortunately, as discussed in this chapter, much of the work in this area has been highly qualitative in nature and does not lead to an unambiguous conclusion.

[50] H. C. Brown and M.-H. Rei, *J. Amer. Chem. Soc.*, **86**, 5008 (1964).
[51] F. Kaplan, P. Cross, and R. Prinstein, *J. Amer. Chem. Soc.*, **92**, 1446 (1970).
[52] R. D. Wietig, R. H. Staley, and J. L. Beauchamp, *J. Amer. Chem. Soc.*, **96**, 7552 (1974).
[53] H. Hogeveen, C. J. Gaasbeek, and A. F. Bickel, *Rec. Trav. Chim.*, **88**, 703 (1969).
[54] H. Hogeveen, F. Baardman, and C. F. Roobeck, *Rec. Trav. Chim.*, **89**, 227 (1970).
[55] E. M. Arnett and C. Petro, private communication.

In any event, it is desirable to retain proper perspective. Numerous cations capable of degenerative rearrangement have now been subjected to nmr examination (Table 13.1). Almost all of them have been assigned equilibrating classical structures. Indeed, only one of the ions in Table 13.1, 2-norbornyl, may exist in σ-bridged form in superacid media, although a sound decision does not appear possible on the basis of the data currently available. Clearly the nonclassical ion problem has by now become shrunken to a size where it may be intellectually interesting, but is no longer of significance to the advance of organic chemistry.

Comments

The historical perspective, as discussed by Brown, is indeed generally valid. Although the possibility of rapidly equilibrating cations had often been considered during the long history of carbocation research, bridged ion alternatives were usually formulated instead in the period after 1945. The ability to observe carbocations directly in superacid media changed this situation rapidly, especially since the observation of temperature-dependent nmr spectra provided direct evidence that rapid equilibration was taking place.

I do not agree, however, "that rapid equilibration is the norm" "under stable ion conditions." Table 13.1 is misleading; there are at least as many carbocations under stable ion conditions which have been assigned nonclassical, rather than rapidly equilibrating classical, structures on the basis of reasonably convincing evidence. After all, one does not need an "immense amount of data" to prove a point; a single definitive experiment should suffice. For example, the δ ^{13}C nonadditivity criterion may well prove to be decisive. The inclusion of stable ions like **30–32** referred to by Brown and **33–39**, *inter alia* would help to restore balance to Table 13.1.

$30^{8,56}$ $31^{9,57}$ 32^{10} 33^{10}

34^{10} $35^{46,58}$ $36^{46,58,59}$ $37^{60,61}$

---

[56]However, G. A. Olah, G. Liang, and S. P. Jindal, *J. Amer. Chem. Soc.*, **98**, 2508 (1976) stress the carbonium ion (classical) character of this species.

[57]D. Lenoir, P. Mison, G. Liang, G. A. Olah, and P. v. R. Schleyer, unpublished observations.

[58]S. Winstein, *Quart. Rev. (London)*, **23**, 1411 (1969), and *Carbonium Ions*, G. A. Olah and P. v. R. Schleyer, Eds., Vol. III, Wiley-Interscience, New York, 1972, Chapter 22; P. R. Story and B. C. Clark, Jr., *ibid.*, Chapter 23.

[59]G. A. Olah and G. Liang, *J. Amer. Chem. Soc.*, **97**, 6803 (1975).

[60]R. M. Coates and E. R. Fretz, *J. Amer. Chem. Soc.*, **97**, 2538 (1975). See comments, Chapter 14.

[61]Review: R. E. Leone, J. C. Barborak and P. v. R. Schleyer, *Carbonium Ions*, Vol. IV, G. A. Olah and P. v. R. Schleyer, Eds., Wiley-Interscience, New York, 1973, Chapter 33.

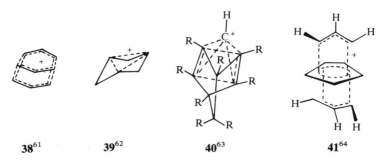

38[61]   **39**[62]   **40**[63]   **41**[64]

Strictly speaking, carbocations are truly "free" only when generated in the isolated state in the gas phase. Stable ions in superacid media must be associated with a counter ion, if only electrostatically at a distance, and must at least benefit from "general" or "bulk" solvation, which may be rather unspecific in nature. However, the relative energies of isomeric carbocations are remarkably similar in the gas phase and in superacid media.[2,65,66] This suggests that isomeric carbocations are solvated to nearly the same total extent irrespective of their classical or bridged nature. Perhaps the detailed structures of these species are also much the same in the gas phase and in solution.

In the more nucleophilic solvolytic media, carbocations clearly have less chance of being "free," and specific interactions with the gegenion and one or more solvent molecules may alter significantly the structure, energies, and behavior of the intermediate. We now recognize that solvent assistance plays a significant role in the solvolysis of most simple primary and many secondary systems; the resulting intermediates are probably best described as having "cationoid" and not "carbenium ion" character.[65] Even so, compounds in which participation can occur are less affected by solvation effects on solvolysis. This is because the neighboring group, rather than the solvent, supplies the nucleophilic "push" needed to help ionize the leaving group. Of course, the neighboring group must be properly oriented. Anchimeric assistance is expected from *exo-*, but not from *endo-*norbornyl solvolysis, even though the structure of the intermediate ion may be the same provided the lifetimes are long enough. In an anchimerically assisted solvolysis, the transition state, although related to the intermediate ion, has not developed as much charge nor has bridging proceeded to the same

[62]S. Masamune, M. Sakai, A. V. Kemp-Jones, and T. Nakashima, *Can. J. Chem.*, **52**, 855 (1974); S. Masamune, M. Sakai, and A. V. Kemp-Jones, *ibid.*, **52**, 858 (1974).

[63]H. Hart and M. Kuzuya, *J. Amer. Chem. Soc.*, **96**, 6436 (1974); S. Masamune, *Pure Appl. Chem.*, **44**, 861 (1975).

[64]M. J. Goldstein and S. A. Kline, *J. Amer. Chem. Soc.*, **95**, 935 (1973).

[65]J. L. Fry, J. M. Harris, R. C. Bingham, and P. v. R. Schleyer, *J. Amer. Chem. Soc.*, **92**, 2540 (1970).

[66]E. W. Bittner, E. M. Arnett, and M. Saunders, *J. Amer. Chem. Soc.*, **98**, 3734 (1976).

degree. "Bridging lags behind ionization"; in a nonclassical transition state, there is more positive charge at the reaction site (C2 of norbornyl) than at a rearranged site (e.g., C1 of norbornyl). Solvolysis and stable-ion studies provide complementary information, which certainly is related, provided carbocation and not S_N2 behavior is being compared.

I do not wish to debate the extent to which ^{13}C chemical shifts measure charge delocalization. I think it is fair to conclude that charge delocalization plays a significant, if not a major, role in determining the ^{13}C chemical shifts of carbocations, but other factors are also involved to an important extent. At the present time, it is not possible to sort out all effects quantitatively.

For present purposes, this is less important than interpretations based on observed deviations of ^{13}C, ^{1}H, ^{19}F, etc. chemical shifts from "normal" behavior. As Olah has repeatedly offered detailed arguments, and analyses supporting his interpretation have been presented in the specialized literature,[38] I will only summarize the situation and state my own conclusions.

Static classical carbocations exhibit ^{13}C chemical shifts for their carbocation centers which are deshielded to remarkable degrees, hundreds of ppm from the usual range (e.g., **13**, **14**, **16–18**). Rapidly equilibrating classical carbocations exhibit chemical shifts which are averaged over the positions undergoing exchange. The same procedure applied to species assigned nonclassical structures fails, often badly. A large part of the deshielding expected from the positive charge seems to disappear.

Ideally, the ^{13}C chemical shifts of both classical and nonclassical forms of an ion should be calculated for comparison with experiment. This is not yet possible. However, there are three lines of evidence which indicate that bridging is responsible for the ^{13}C chemical shift nonadditivity.

Ditchfield and Miller[67] have calculated the ^{13}C chemical shifts of the classical and of the bridged ethyl cation. The average δ ^{13}C of the former, which would represent the value to be expected for a rapidly equilibrating set of classical ethyl cations, is deshielded *by 50 ppm* relative to δ ^{13}C of bridged $C_2H_5^+$.

Coates' cation (**37**, R = H)[60,61] is now conceded to be a nonclassical species. The ions **32**[10], **35**[46,58], and **36**[46,58,59] appear to me to be equally secure. They exhibit the same type of deviations from chemical shift additivity as does 2-norbornyl, and the same "loss" of deshielding.

Finally, a parallel between the ^{13}C chemical shifts in carbocations and the ^{11}B chemical shifts in the corresponding boranes has been noted.[68] This parallel has been employed recently by R. E. Williams[69] to estimate the

[67]R. Ditchfield and D. D. Miller, *J. Amer. Chem. Soc.*, **93**, 5287 (1971).

[68]R. Nöth and B. Wrackmeyer, *Chem. Ber.*, **107**, 3089 (1974); B. F. Spielvogel, W. R. Nutt, and R. A. Izydore, *J. Amer. Chem. Soc.*, **97**, 1609 (1975).

[69]R. E. Williams, paper submitted for presentation at the Third International Meeting on Boron Chemistry, Ettal, Germany, July, 1976, Abstract No. 31; private communication.

chemical shifts of nonclassical carbocations like **32** with success. It has even been applied to the norbornyl problem, but details of this work have not yet appeared. The ^{13}C chemical shifts of the pentacoordinate and hexacoordinate carbons in carboranes should also be instructive. Such analyses are now underway.[70]

Olah and Mateescu are to be congratulated for their boldness, initiative, and perseverance in applying ESCA spectroscopy to the problem of carbocation structures.[4,44] The results should indeed provide unambiguous answers. The ESCA spectra of a number of carbocations have been published by the Cleveland group, but only the conclusions concerning the 2-norbornyl cation have been questioned. For example, the cyclopentyl cation exhibits only single ^{1}H and ^{13}C signals down to very low temperatures, but its ESCA spectrum exhibits a 4:1 pattern consistent with a classical ion. Is not this result to be believed?

Fortunately, it appears to be only a matter of time before the matter will be resolved. The ESCA spectrum of the 2-norbornyl cation is being repeated in a vacuum spectrometer, which should obviate the possible problems of surface contamination.[71a] Sophisticated "hole state" calculations have been carried out which should be able to predict the ESCA spectra of classical and nonclassical alternatives accurately.[71a] The results conform closely to Olah's experimental spectra (but not Dewar's MINDO/3 calculations),[71b] and confirm the conclusion that the norbornyl cation is bridged.

A number of recent investigations have provided measurements of the stability of the 2-norbornyl cation in the gas phase and in superacid solution.[51-55,66,72-75] The techniques employed have varied considerably, but not the conclusions, that is, the 2-norbornyl cations is 6–10 kcal/mole more stable than expected of a classical secondary species!

On the basis of an ion cyclotron resonance study, "the bicyclo[2.2.1]heptyl cation compared to its hydrocarbon is approximately 6 kcal/mole more stable than the bicyclo[2.2.2]octyl cation compared to its hydrocarbon."[51] Similar results were obtained by Beauchamp's group.[52]

The reversible carbonylation of the 2-norbornyl cation in $FHSO_3$-SbF_5 gave a complete free-energy diagram.[72] "These data show that

---

[70]G. A. Olah and G. Liang, private communication.

[71a]D. T. Clark, private communication; D. T. Clark, B. J. Cromarty, and L. Colling, *J. Amer. Chem. Soc.*, submitted. Also see D. W. Goetz and L. C. Allen, *ibid.*, submitted.

[71b]M. J. S. Dewar, R. C. Haddon, A. Komornicki, and H. Rzepa, *J. Amer. Chem. Soc.*, **99**, 377 (1977).

[72]H. Hogeveen, *Advan. Phys. Org. Chem.*, **10**, 29 (1973).

[73]See T. S. Sorensen, *Accounts Chem. Res.* **9**, 257 (1976).

[74]J. J. Solomon and F. H. Field, *J. Amer. Chem. Soc.*, **98**, 1567 (1976).

the 2-norbornyl ion is . . . about $8 \, \text{kcal mole}^{-1}$ more stabilized than *secondary* alkyl cations."

Sorensen and his coworkers[73] have estimated the relative stabilities of various 2-norbornyl cations based on rearrangement rates and equilibria in superacid media. The secondary 2-norbornyl cation is only 7.5 kcal/mole less stable than the tertiary 2-methyl-2-norbornyl cation. This value is much lower than the usual secondary-tertiary energy difference found in classical carbocations, e.g., the 14.5 ± 0.5 measured directly for the 2-butyl $\rightarrow$ t-butyl cation rearrangement.[64] Such comparisons indicate the 2-norbornyl cation to be stabilized by 7 kcal/mole. Direct measurement of heats of ionization (6), now further refined,[55] gives a similar value and essentially eliminates the discrepancy implied on p. 253.

Time-resolved high-pressure mass spectrometry results led Solomon and Field[74] to conclude, ". . . the norbornyl cation has 10 kcal/mole more stability than that which would be expected for a secondary species." The 2-norbornyl heat of formation has now been confirmed by measurements of the proton affinity of norbornene.[75]

All of these investigations, it is true, do not prove the 2-norbornyl cation to be bridged, but that seems to be the most reasonable conclusion.

There is, in fact, no real discrepancy between gas-phase and solution data for the relative stabilities of the t-butyl and the 2-norbornyl cations. In the gas phase, ions are stabilized internally, and size effects are important.[52] Thus, 2-norbornyl benefits, relative to t-butyl, by having seven carbons vs. four. A more appropriate comparison would be t-heptyl[76] vs. 2-norbornyl. In condensed phases, such "internal solvation" is less important, since bulk solvation attenuates or eliminates differences produced by size. The remarkable thing is that *secondary* 2-norbornyl is so close in stability to *tertiary*-butyl when large *secondary–tertiary* differences are the rule with classical carbocations.

Nonclassical carbonium ions are not as prevalent as once believed. But whether their actual number is large or small, Professor Brown and I agree that the phenomenon they illustrate is of great significance to organic chemistry. As discussed in my Comments to Chapter 1, multicenter electron-deficient bonding involving carbon is commonly encountered in other areas; it is important to recognize, to understand, and to utilize all the diverse structural arrangements chemistry affords.

[75]R. H. Staley, R. D. Weiting, and J. L. Beauchamp, *J. Amer. Chem. Soc.*, submitted; J. L. Beauchamp, private communication.

[76]F. P. Lossing and A. Maccoll, *Can. J. Chem.*, **54**, 990 (1976). This paper shows that the tertiary–secondary carbocation energy difference is normally 17 kcal/mole in the gas phase, in contrast to the behavior of norbornyl.[74]

14

New Concepts
—New Systems

14.1. Introduction

The extraordinary difficulties that we and others have encountered in obtaining unambiguous evidence for the long postulated σ-participation in *exo*-norbornyl derivatives with accompanying charge delocalization from the 2- to the 1- and 6-positions[1] have been reviewed in this book. These difficulties have led to new proposals for an alternative stereoelectronic effect to favor the ionization of the *exo* isomer, but without the σ-bridge of the original proposal. These new proposals will now be considered. In addition, new systems have been advanced as providing less ambiguous examples for σ-bridging. Some of these will be briefly considered here.

14.2. The Hyperconjugative Model

In the hyperconjugative model[2-4] the C1–C6 bonding pair of the 2-norbornyl structure provides an electronic contribution to facilitate ionization (1) of the *exo* derivative (**1**) without σ-bridging or major distortion of the structure of the cation (**2**).

$$\text{1} \quad \longrightarrow \quad \text{2} \quad + \quad X^- \tag{1}$$

[1]S. Winstein and D. Trifan, *J. Amer. Chem. Soc.*, **74**, 1147, 1154 (1952).
[2]E. Kosower, *Physical Organic Chemistry*, Wiley, New York, 1968, pp. 136–172.
[3]F. R. Jensen and B. E. Smart, *J. Amer. Chem. Soc.*, **91**, 5688 (1969).
[4]T. G. Traylor, W. Hanstein, H. J. Berwin, N. A. Clinton, and R. S. Brown, *J. Amer. Chem. Soc.*, **93**, 5715 (1971).

The proposal is a reasonable one on the basis of the stereochemical relationship of the developing *p*-orbital at C2 (**3**) to the C6–C1 bonding pair:

$$\text{(2)}$$

This *endo* lobe of the *p*-orbital at the developing cationic center is essentially parallel to the C6–C1 bond. Consequently, it is in a favorable position for conjugative interaction with this bond. σ-Participation would require a major distortion of the structure to place this lobe in a position to interact end-on with the C1–C6 bonding pair.

Jensen and Smart suggested that C–C hyperconjugation in the 2-norbornyl cation might be depicted as involving the canonical structures **2′** ↔ **2″** in valence-bond terms[3]:

However, they believed that these canonical forms involving σ-bond breaking may be misleading when applied to hyperconjugation. They proposed that hyperconjugation should be considered as a phenomenon by which σ-bonding electrons are only partially delocalized to an electron-deficient center and that these "participating" electrons formally remain in the σ-bond. In other words, the movement of electrons and atoms is only slight, but enough to provide stabilization at an adjacent developing electron-deficient center.

A similar interpretation has been advanced by Traylor.[4] He proposes the term "vertical stabilization" for the electronic stabilization of carbonium ions by delocalization of neighboring σ bonds without significant changes in the geometry of the cation, such as 2-norbornyl (**2**).

It is not clear as to how this model differs from the simple classical cation. In the classical picture (Chapter 6), the 2-norbornyl cation would be stabilized by electronic shifts from the carbon structure as described by the terms, "inductive," "field," and "hyperconjugative." In other words, there would be an electronic interaction which would distribute the charge throughout the system.

The hyperconjugative model proposed by Jensen and by Traylor resembles the classical ion in not having a plane of symmetry. Consequently, it is also necessary here to postulate equilibration of a pair of isomeric cations at a rate that is rapid compared to their capture by solvent.

The new feature that is introduced is the proposal that the proposed hyperconjugative or vertical stabilization is far more effective for the *exo* isomer (**1**) than for the *endo* isomer (**4**). It is considered that the C1–C7 bonding pair is in a less favorable stereochemical position (**5**) to interact with the developing *exo* lobe (**3**).

$$\text{(3)}$$

Fortunately, this is a point which appears to be susceptible to experimental test (Section 14.4).

14.3. The Exo-6-H Participation Model

Still another, quite different model for 2-norbornyl has been advanced. Olah was impressed with the observation that in the 2-methyl-2-norbornyl cation the *exo*-6-H resonance appears at 3.28 ppm, in contrast to a value of 1.09 ppm for *endo*-6-H.[5] He attributed the difference in the pmr resonances to a selective withdrawing of electron density from the *exo*-C6–H bond. He proposed that the *endo* lobe of the developing *p*-orbital at C2 interacts with the back lobe of the *exo*-C6–H orbital (**6**) and not with the C1–C6 bond (**4**).[5,6]

$$\text{(4)}$$

Again the proposal appears to be susceptible to experimental test (Section 14.4). Can we find a preference for electron supply from the *exo* direction as compared to the *endo*?

[5]G. A. Olah, A. M. White, J. R. De Member, A. Commeyras, and C. Y. Lui, *J. Amer. Chem. Soc.*, **92**, 4627 (1970).
[6]G. A. Olah, J. R. De Member, C. Y. Lui, and R. D. Porter, *J. Amer. Chem. Soc.*, **93**, 1442 (1971).

14.4. The Search for a Stereoelectronic Contribution

The hyperconjugative model proposed by Jensen[3] and by Traylor[4] and the *exo*-6–H participation model of Olah[5,6] account for the high *exo*:*endo* rate ratio exhibited in the solvolysis of 2-norbornyl derivatives[1] to an electronic contribution which stabilizes the *exo* transition state more than the *endo*. The interpretation differs from the older nonclassical ion proposal in that σ-participation and σ-bridging are not involved in the transition state and major distortion of the structure is not essential for the operation of the electronic contribution facilitating ionization of the *exo* isomer.

We are still faced with the same problem. Can we devise experiments which provide independent support for the proposal of electronic contributions to a developing electron deficiency at C2 that is markedly greater from the *exo* direction than from the *endo*?

Jensen and Smart examined the aluminum chloride-catalyzed benzoylation of the phenylnorbornanes in ethylene dichloride at 25°C as a probe for enhanced electronic contributions from the norbornyl system.[7] The simple monoalkylbenzenes exhibit a large partial rate factor for substitution in the *para* position, p_f, decreasing with increasing branching of the alkyl group (5).[8]

$$p_f \ (25°C): \quad 633 \qquad 563 \qquad 519 \qquad 398$$

(5)

The enhanced rate of electrophilic substitution *para* to such alkyl groups, as compared to hydrogen, is attributed to hyperconjugation. The decrease in rate as the alkyl group becomes more branched has been attributed to the lower effectiveness of C–C as compared to C–H hyperconjugation.[9–11] Others have proposed that the rate sequence, Me > Et > i-Pr > t-Bu is due to steric hindrance of solvation.[12,13]

[7]F. R. Jensen and B. E. Smart, *J. Amer. Chem. Soc.*, **91**, 5686 (1969).

[8]H. C. Brown and G. Marino, *J. Amer. Chem. Soc.*, **81**, 5611 (1959); F. R. Jensen, G. Marino, and H. C. Brown, *ibid.*, **81**, 3303 (1959).

[9]R. S. Mulliken, C. A. Rieke, and W. G. Brown, *J. Amer. Chem. Soc.*, **63**, 41 (1941).

[10]E. Berliner and F. J. Bondhus, *J. Amer. Chem. Soc.*, **70**, 854 (1948).

[11]H. C. Brown, J. D. Brady, M. Grayson, and W. H. Bonner, *J. Amer. Chem. Soc.*, **79**, 1897 (1957).

[12]W. M. Schubert, J. M. Craven, R. G. Minton, and R. B. Murphy, *Tetrahedron*, **5**, 194 (1959).

[13]E. M. Arnett and J. M. Abbound, *J. Amer. Chem. Soc.*, **97**, 3865 (1975); W. J. Hehre, R. T. McIver, Jr., J. A. Pople, and P. v. R. Schleyer, *ibid.*, **96**, 7162 (1974).

On the basis of either of these two explanations, one would expect toluene to undergo benzoylation faster than any of the phenylnorbornanes. However, the opposite is true. Indeed, the compound exhibiting the fastest rate is 1-phenylnorbornane (**7**), which contains no α-C–H bonds (6).

	7	8	9	10	(6)
p_f (25°C):	1790	822	1630	1040	

In fact, 1-phenylnorbornane exhibits a p_f factor some 4.5 times that of 1-*tert*-butylbenzene. This supports the conclusion that strained C–C bonds, such as are present in norbornane, can exhibit an enhanced ability to stabilize an electron-deficient center.

The *exo* isomer **9** exhibits a rate significantly higher than that of the *endo* derivative **10**. Indeed, the p_f values for the 7- (**8**), *exo*-2 (**9**) and *endo*-2 (**10**) exhibit an order which varies qualitatively with the observed rates of acetolysis of the corresponding tosylates (7).

			(7)
k (25°C):	6.36×10^{-15}	2.33×10^{-5}	8.28×10^{-8}
RR (25°C):	7.68×10^{-8}	280	1.00

The differences in the p_f factors are small, much smaller than the differences in the observed rates of acetolysis. However, this could be a consequence of the fact that in solvolysis the electron deficiency is developed directly on one of the ring atoms, whereas in benzoylation the charge is delocalized throughout the aromatic ring and only a small portion of the electron deficiency reaches the carbon atom bonded to the norbornane structure.

More serious is the absence of any proportional effect. The p_f value for 7-phenylnorbornane (**8**) is only modestly smaller than that for *endo*-2-phenylnorbornane (**10**), in spite of the huge differences in the rates of solvolysis.

Nevertheless, these results are important—they provide the first independent evidence for an electronic effect which is larger from the *exo* direction than from the *endo* direction (compare **9** with **10**).

The benzoylation reaction possesses both favorable and unfavorable characteristics for such a study of electronic contribution.[8] On the favorable side is the high electron demand, with $\rho \approx ca - 9.0$, and predominant substitution (~95%) in the position *para* to an alkyl substituent, presumably resulting from a large steric factor for the reagent. On the negative side, there is the possibility that the large steric requirements could influence in a modest way the rates of substitution of the *exo*- and *endo*-phenyl groups. In other words, it is conceivable that the factor of <2 observed for **9** and **10** could arise from a small steric effect influencing the rate of benzoylation of the *endo*-2-phenyl group.

The solvolysis of *tert*-cumyl chlorides appears to be remarkably free of complications and it has been applied as a probe to a large number of systems without the appearance of any significant complications.[14] Consequently, this probe was adopted to examine the relative effectiveness of electronic contributions from *exo*- and *endo*-norbornyl groups in the *para* position of the *tert*-cumyl system (**11**) in stabilizing the electron deficiency in the developing cation.[15]

The enhanced ability of a *p*-cyclopropyl ring (**13**) to supply electrons to an electron-deficient center as compared to an isopropyl group (**12**) is unambiguously indicated (8).[16]

| RR (25°C): | 1.00 | 18.8 | 154 |

On the other hand, a *p*-cyclobutyl group (**14**) does not cause any unusual rate acceleration compared to *p*-isopropyl (**12**), *p*-cyclopentyl (**15**), or *p*-cyclohexyl (**16**) (9).[17]

[14]H. C. Brown, *Steric Effects in Conjugated Systems*, G. W. Gray, Ed., Butterworth, London, 1968, pp. 100–118.

[15]H. C. Brown, B. G. Gnedin, K. Takeuchi, and E. N. Peters, *J. Amer. Chem. Sóc.*, **97**, 610 (1975).

[16]H. C. Brown and J. D. Cleveland, *J. Org. Chem.*, **41**, 1792 (1976).

[17]R. C. Hahn, T. F. Corbin, and H. Shechter, *J. Amer. Chem. Soc.*, **90**, 3404 (1968).

$$
\begin{array}{ccc}
\underset{14}{\text{H}_3\text{C}-\overset{\overset{\displaystyle \text{Cl}}{|}}{\text{C}}-\text{CH}_3} &
\underset{15}{\text{H}_3\text{C}-\overset{\overset{\displaystyle \text{Cl}}{|}}{\text{C}}-\text{CH}_3} &
\underset{16}{\text{H}_3\text{C}-\overset{\overset{\displaystyle \text{Cl}}{|}}{\text{C}}-\text{CH}_3}
\end{array}
\tag{9}
$$

RR (25°C):   20.7   23.7   19.6

The solvolysis results reveal that *exo*-norbornyl (**17**) is slightly better than *endo*-norbornyl (**18**) in facilitating the solvolysis (10).[15]

$$
\begin{array}{cc}
\underset{17}{\text{H}_3\text{C}-\overset{\overset{\displaystyle \text{Cl}}{|}}{\text{C}}-\text{CH}_3} &
\underset{18}{\text{H}_3\text{C}-\overset{\overset{\displaystyle \text{Cl}}{|}}{\text{C}}-\text{CH}_3}
\end{array}
\tag{10}
$$

RR (25°C):   25.2   21.8

These results are summarized in Table 14.1.

The difference in rates is small, smaller than that observed in the benzoylation reaction (6) and even smaller than that anticipated from the difference in the ρ values (benzoylation, $\rho \approx -9$,[8] solvolysis of *tert*-cumyl

TABLE 14.1. *Rates of Solvolysis of p-Alkyl-tert-cumyl Chlorides in 90% Acetone at 25°C*[15]

Alkyl group	$10^6 k_1$, sec^{-1} at 25°C	$\Delta H^{\ddagger}$, kcal mol^{-1}	$\Delta S^{\ddagger}$, eu	RR at 25°C	σ^+
Hydrogen	1.24	18.8	−12.4	1.00	(0.00)
Isopropyl	23.3	17.4	−12.2	18.8	−0.280
exo-2-Norbornyl	31.2	17.3	−11.8	25.2	−0.309
endo-2-Norbornyl	27.0	17.9	−10.2	21.8	−0.295
Cyclopentyl	29.4	17.4	−11.6	23.7	−0.302
Cyclohexyl	24.3	17.6	−11.6	19.6	−0.285
Cyclopropyl	190.5	16.0	−12.6	154	−0.462

chloride, $\rho = -4.54$[18]). Indeed, the difference in rates is comparable to those observed for **12, 14, 15,** and **16**.

Perhaps it would be possible to enhance the effect by placing the developing positive charge α to the ring system. In this position, the developing electron deficiency should make a much greater electronic demand on the system for stabilization. That this is indeed the case is clearly revealed by comparing the relative effects of isopropyl and cyclopropyl in **12** and **13**, respectively, with their effects in **19** and **20** (11).[19]

$$
\begin{array}{cccc}
\text{Cl} & \text{Cl} & \text{OPNB} & \text{OPNB} \\
| & | & | & | \\
\text{H}_3\text{C}-\text{C}-\text{CH}_3 & \text{H}_3\text{C}-\text{C}-\text{CH}_3 & \text{Ph}-\text{C}-\text{CH}_3 & \text{Ph}-\text{C}-\text{CH}_3
\end{array}
$$

12   **13**   **19**   **20**

(11)

RR (25°C): 1.00   8.2   1.00   25,300

However, it is the *endo* isomer **22**, not the *exo* isomer **21**, that exhibits the faster rate (12).[20] Presumably, the faster rate of the *endo* isomer is the result of relief of steric strain (Section 2.2) and it swamps any enhanced electronic contribution in the *exo* isomer.

21   **22**

(12)

RR (25°C):   0.055   1.0

The tool of increasing electron demand (Chapter 10) should be capable of detecting such enhanced electron supply in the *exo* isomer. However, the $\rho+$ values for the *exo* (**23**) and *endo* (**24**) derivatives are indistinguishable (13).[20]

23   **24**

(13)

$\rho+$ -4.44 -4.47

[18]H. C. Brown and Y. Okamoto, *J. Amer. Chem. Soc.*, **80**, 4979 (1958).
[19]E. N. Peters and H. C. Brown, *J. Amer. Chem. Soc.*, **95**, 2397 (1973).
[20]H. C. Brown and M. Ravindranathan, manuscript in preparation.

TABLE 14.2. *Effect of Increasing Electron Demand in exo- and endo-Norbornyl Derivatives*[20]

OPNB	$10^6 k_1$, sec^{-1} at 25°C	$\Delta H^{\ddagger}$, kcal mol^{-1}	$\Delta S^{\ddagger}$, eu	RR at 25°C
23 p-CH$_3$O	112			0.16
p-H	0.0472	25.8	−5.7	0.33
p-CF$_3$	6.73×10^{-5}	31.2	−0.6	0.22
3,5-(CF$_3$)$_2$	1.04×10^{-6}	34.8	2.4	0.19
24 p-CH$_3$O	708			1.00
p-H	0.144	25.7	−8.6	1.00
p-CF$_3$	3.02×10^{-4}	29.4	−3.4	1.00
3,5-(CF$_3$)$_2$	5.52×10^{-6}	32.5	−0.9	1.00

These data are summarized in Table 14.2.

Finally, even when the developing electron deficiency is actually on C2 (Section 10.5), no differential electron supply in the *exo* isomer (**26**) as compared with the *endo* isomer (**27**) is observed (14).[21]

$$\rho+: \qquad -3.82 \qquad\qquad -3.82 \qquad\qquad -3.72 \tag{14}$$

If we take $\rho+$ as a measure of the relative effectiveness of the electron supply from alkyl or alicyclic group (15), then it would appear that the cyclopentyl, *exo*-norbornyl, and *endo*-norbornyl are better for such supply than methyl (**28**), isopropyl (**29**), cyclobutyl (**30**), or cyclohexyl (**31**). However, for the objectives of the present study, it is the similarity in the $\rho+$ values for the *exo*(**26**) and *endo* (**27**) derivatives that would appear to be the significant result.

$$\rho+: \qquad -4.72 \qquad\qquad -4.76 \qquad\qquad -4.91 \qquad -4.60 \tag{15}$$

In this connection it should be pointed out that Battiste and Fiato have recently proposed that in contrast to the original position[1] the rate of the *exo*

[21]H. C. Brown, M. Ravindranathan, K. Takeuchi, and E. N. Peters, *J. Amer. Chem. Soc.*, **97**, 2899 (1975).

should not be compared with that of the *endo*.[22] Nor do they wish to use the cyclopentyl derivative (**25**) as the reference system. Instead, they argue for adoption of the isopropyl system (**29**) as the reference.

	29	**26**	**27**	
p-CH₃O	1.00	175	0.62	(16)
p-H	1.00	795	5.57	
p-CF₃	1.00	3140	16.8	
3,5-(CF₃)₂	1.00	8460	48.1	

Adoption of this system as the reference indicates equivalent electron supply with increasing electron demand in both the *exo* ($\Delta\rho = 0.94$) and the *endo* ($\Delta\rho = 1.04$) system (16). With the increase in electron demand represented by the change in substituents from p-CH₃O to 3,5-(CF₃)₂, the relative reactivities increase by a factor of 78 in the *endo* isomers (**27**) and by a somewhat smaller factor, 48, in the *exo*. Again, even this approach fails to reveal an enhanced electron supply by whatever mechanism in the *exo* system.

Finally, let us consider the possibility that electron demand in these tertiary systems is small compared to the demand in secondary systems.[23] Solvolyses are proposed to be limiting in trifluoroacetic acid.[24] Moreover this solvent has the highest ionizing power ($Y = 4.57$) and the lowest nucleophilicity ($N = -5.56$) of the solvents utilized in solvolysis studies.[24] Consequently, solvolysis in this system should make an exceptionally powerful demand on the norbornyl system (or other secondary derivatives) for electron supply.

Evidence that this is the case is indicated by the relative rates in trifluoroacetic acid (17).[25] It should be noted that the increase in rate from isopropyl to cyclopentyl (127.5) is equivalent to the increase from cyclopen-

RR (25°C):	1.00	127.5	21900	(17)

[22]M. A. Battiste and R. A. Fiato, *Tetrahedron Lett.*, 1255 (1975).

[23]G. D. Sargent, *Carbonium Ions*, Vol. III, G. A. Olah and P. v. R. Schleyer, Eds., Wiley–Interscience, New York, 1972, Chapter 24.

[24]F. L. Schadt, P. v. R. Schleyer, and T. W. Bentley, *Tetrahedron Lett.*, 2335 (1974).

[25]J. E. Nordlander, R. R. Gruetzmacher, W. J. Kelly, and S. P. Jindal, *J. Amer. Chem. Soc.*, **96**, 181 (1974).

TABLE 14.3. *Methyl–Hydrogen Rate Ratios*

System	Rate constants		
	s-ROTs in CF$_3$CO$_2$H	*t*-ROPNB in 80% acetone	$\dfrac{\text{Methyl}}{\text{Hydrogen}} \times 10^6$
Isopropyl	2.14×10^{-5}	7.45×10^{-11}	3.48/c
Cyclopentyl	2.73×10^{-3}	2.11×10^{-9}	0.77/c
Cyclohexyl	2.7×10^{-4}	5.48×10^{-11}	0.20/c
exo-Norbornyl	0.468	1.0×10^{-8}	0.021/c
endo-Norbornyl	4.17×10^{-4}	1.13×10^{-11}	0.027/c
2-Adamantyl	9.0×10^{-4}	1.43×10^{-10}	0.16/c

tyl to *exo*-norbornyl (171). It has been proposed that these results can be interpreted to reveal that 50% of the *exo*:*endo* rate ratio is the result of enhanced electronic contributions in the *exo* isomer and 50% the result of steric retardation of ionization in the *endo* isomer.[26]

However, let us calculate the Me/H ratio (Section 11.5) for the *exo* and *endo* derivatives to see if the data reveal an electronic contribution in the *exo* isomer which is absent in the *endo* isomer. Let us assume that there is a constant factor *c* which we can use to correct from tosylate to *p*-nitrobenzoate as the leaving group and from trifluoroacetic acid to 80% acetone as the solvent. Then the calculated rate for the secondary can be compared directly with the rate of the corresponding tertiary *p*-nitrobenzoate in 80% acetone (18).

$$k_{Me}/k_H = \frac{k \left(\begin{array}{c} \text{Me} \\ \diagdown\diagup \\ \text{C} \\ \diagup\diagdown \\ \text{OPNB} \end{array} \right)}{k \left(\begin{array}{c} \text{H} \\ \diagdown\diagup \\ \text{C} \\ \diagup\diagdown \\ \text{OTs} \end{array} \right) c} \tag{18}$$

The data are summarized in Table 14.3.

The decreased value for Me/H for *exo*-norbornyl utilizing the data for the secondary derivative is indeed consistent with an enhanced electronic contribution from the *exo*-norbornyl structure under the huge electronic demand of these conditions. However, the same decrease is observed in *endo*-norbornyl! We can only conclude that enhanced electron supply is occurring in both the *exo* and *endo* isomers.

[26]T. W. Bentley, *Ann. Rept. Chem. Soc. B*, 119 (1974).

As of the time this is written, there does not appear to be available any unambiguous evidence for enhanced electron supply in the *exo* isomer over and above that available in the *endo* isomer.

14.5. New Proposed σ-Bridged Systems

At the time our research program on nonclassical ions was initiated there was general agreement that cyclopropylcarbinyl and 2-norbornyl provided the best possible examples for σ-bridged cations. In this book we have examined critically the evidence made available by my group and by others. We have failed to find evidence to support the σ-bridged interpretation.

In the meantime, a number of other systems have been proposed as providing new examples for such bridging. These systems may provide authentic examples of such σ-bridged cations. However, they have not yet been subjected to critical examination. In view of the special objectives of the present book, they will be discussed here only briefly.

The strained carbon–carbon bonds of the cyclopropane ring provide a transition from the electron-rich carbon–carbon double bonds and the much more tightly held σ-electrons of the usual σ bond. Indeed, there is evidence that a properly situated cyclopropyl ring (**33**) can provide πσ-participation as large as or larger than the π-participation provided by a similarly situated double bond (**32**) (19).[27]

| RR (25°C): | 1.00 | 10^{11} | $\sim 10^{14}$ |

This conclusion is supported by a study of the 7-*p*-anisyl groups (20).[28] Whereas a *p*-anisyl group in the 7-position is capable of swamping the

| RR (25°C): | 1.00 | 4×10^3 |

[27]J. S. Haywood-Farmer and R. E. Pincock, *J. Amer. Chem. Soc.*, **91**, 3020 (1969).
[28]P. G. Gassman and A. F. Fentiman, Jr., *J. Amer. Chem. Soc.*, **92**, 2551 (1970).

π-participation by the double bond in **32**, it does not entirely eliminate $\pi\sigma$-participation in **34**.

The fact that we observe participation in **33** and **34**, does not mean that we are observing the formation of nonclassical cation intermediates **36** and **37**. In both **33** and **34** the solvolysis gives rearranged products **35** ($R = H$ or p-An).

36　　　　**37**

A better case for a true $\pi\sigma$-bridged intermediate is provided by the acetolysis of the 3-bicyclo[3.1.0]hexyl tosylates (**38, 39**).[29] The rate data

38　　　　**39**　　　　**40**

alone fail to exhibit the enhancement to be anticipated for the formation of a stabilized transition state leading to the formation of a stabilized nonclassical intermediate (**40**), as revealed by the relative rates shown in (21).[29]

RR (50°C):　　1.00　　　　1/8.3　　　　1/1.4

The failure to observe an increase in **42** over **41** led to the conclusion that π-participation is not significant in **42**.[29,30] It is evident that **38** also fails to show a really significant rate increase.

On the other hand, the products do support some sort of involvement of the $\pi\sigma$-electrons. Thus the *cis* tosylate **38** undergoes acetolysis to give the *cis* acetate quantitatively, whereas the *trans* tosylate **39** gives a mixture of olefin (33%) and *cis* acetate (67%). Possibly the problem is solvent participation in the solvolysis of these secondary tosylates. Perhaps a determination of their solvolysis in a nonparticipating solvent, such as hexafluoroisopropyl alcohol,[24] will reveal a much larger rate acceleration.

It has been reported that the nonclassical ion **40** has been prepared under stable ion conditions.[31] However, we have seen the difficulties in

[29]S. Winstein and J. Sonnenberg, *J. Amer. Chem. Soc.*, **83**, 3235 (1961).
[30]P. D. Bartlett and M. R. Rice, *J. Org. Chem.*, **28**, 3351 (1963).
[31]S. Masamune, M. Sakai, A. V. Kemp-Jones, and T. Nakashima, *Can. J. Chem.*, **52**, 855 (1974).

differentiating between a set of rapidly equilibrating classical ions and the nonclassical structure for ions under stable ion conditions (Chapter 13).

Perhaps the most persuasive case for a $\pi\sigma$-bridged nonclassical ion is that proposed by Coates.[32,33] He established that solvolysis of **43** proceeds with degenerative rearrangement (22), involving either the nonclassical ion **44** or a set of rapidly equilibrating classical ions or ion pairs (**45**).

$$ \qquad \qquad \qquad \qquad \qquad \qquad \qquad \qquad \qquad \qquad \qquad \qquad (22) $$

      43   **44**   **45**

The deuterium derivative **46** gives rise to alcohols (**47**) on solvolysis in which the deuterium is equally distributed among three equivalent positions (23).

$$ \qquad \qquad \qquad \qquad \qquad \qquad \qquad \qquad \qquad \qquad \qquad \qquad (23) $$

      46   **47**

If 7-norbornyl can be considered to be a suitable model for **43**, then the rate of solvolysis is enhanced by a factor of $\sim 10^{12}$. On this basis, the proposal that the solvolysis proceeds through the nonclassical ion **44** appears reasonable. However, it is hoped that some of the criteria applied to the 2-norbornyl system, such as the tool of increasing electron demand and attempts to trap an unsymmetrical cation will be applied to this promising system.

Nonclassical structures have been proposed for a number of pyramidal mono- and dications (**48**),[34] the 2-bicyclo[2.1.1]hexyl cation (**49**)[35] and the 1-methyladamantyl cation (**50**).[36] However, a critical examination of the

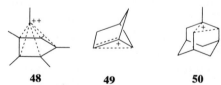

      48   **49**   **50**

[32]R. M. Coates and J. L. Kirkpatrick, *J. Amer. Chem. Soc.*, **92**, 4883 (1970).
[33]R. M. Coates and E. R. Fretz, *J. Amer. Chem. Soc.*, **97**, 2538 (1975).
[34]H. Hogeveen and P. W. Kwant, *Accounts Chem. Res.*, **8**, 413 (1975).
[35]G. Seybold, P. Vogel, M. Saunders, and K. B. Wiberg, *J. Amer. Chem. Soc.*, **95**, 2045 (1973).

evidence reveals that the data are equally consistent with a related set of rapidly equilibrating corresponding classical structures (**51–53**).

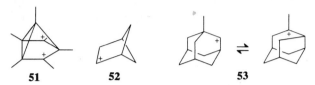

It should be recalled that the first pyramidal structure to be proposed was that for the cation from the solvolysis of 7-chloronorbornadiene[37] (see Figure 1.2), now no longer accepted. The chief argument for **48** is again a discrepancy between the observed ^{13}C shifts for the cation under stable ion conditions and those calculated for a set of five equivalent equilibrating classical cations **51**. We have previously pointed out the uncertainties of this approach (Section 13.5), as evidenced by the fact that it led to the proposal for a σ-bridged cyclopropylcarbinyl cation,[38] in spite of the considerable evidence favoring the cation in the bisected conformation (Chapter 5). The authors indicate that they are attempting to obtain the ESCA spectrum, but are encountering technical difficulties related to those discussed for the 2-norbornyl cation (Section 13.7). Consequently, judgment should be reserved until more definitive evidence becomes available.

The nature of the 2-bicyclo[2.1.1]hexyl cation (**52**) is of special interest in view of its relationship to the 2-norbornyl cation. Does the additional strain in the cyclobutane moiety enhance σ-participation and make this a true nonclassical ion (**49**)? At temperatures between -90 and $-130°C$ the nmr spectra reveal that the six methylene protons become equivalent. It is estimated that the activation enthalpy for the methylene group migration must be less than 7 kcal mol^{-1} and the activation enthalpy for hydride shift must be greater than 13 kcal mol^{-1}. It is considered that the results are more consistent with a set of three rapidly equilibrating σ-bridged cations (**49**) than with a set of six rapidly equilibrating classical cations (**52**).

If the strained cyclobutane moiety facilitates σ-participation, as proposed, it should be possible to detect it in the observed rates of solvolysis. Surprisingly, 2-bicyclo[2.1.1]hexyl tosylate fails to exhibit an enhanced rate. The rate of acetolysis is reported to be 1.7×10^{-5} sec^{-1} at 75°C.[39] This is 3.5 times slower than *endo*-norbornyl tosylate and 1000 times slower than *exo*-norbornyl tosylate to which it is structurally related. With $\nu_{CO} =$

[36]D. Lenoir, D. J. Raber, and P. v. R. Schleyer, *J. Amer. Chem. Soc.*, **96**, 2149 (1974).

[37]S. Winstein and C. Ordronneau, *J. Amer. Chem. Soc.*, **82**, 2084 (1960).

[38]G. A. Olah, D. P. Kelly, C. L. Jeuell, and R. D. Porter, *J. Amer. Chem. Soc.*, **92**, 2544 (1970).

[39]J. Meinwald, Abstracts, 18th National Organic Chemistry Symposium at Columbus, Ohio, 1963, p. 39.

$1764 \, cm^{-1}$ [40] application of the Foote–Schleyer correlation does not reveal an enhanced rate attributable to significant σ-participation. Indeed, Olah has recently reached the conclusion, based on ^{13}C shifts observed for the cation under stable ion conditions, that the bicyclo[2.1.1]hexyl cation is a rapidly equilibrating set of classical cations.[41] If this conclusion is valid, it would render questionable the argument that the slow $3:2$ hydride shifts observed in the 2-norbornyl and in bicyclo[2.1.1]hexyl cations is a valid argument for the σ-bridged structures.[35]

Finally, Daniel Farcasiu, working in the laboratories of Professor Paul von R. Schleyer, extended the study of the 1-methyl-2-adamantyl tosylate[36] to the corresponding 1-cyano- and 1-carboethoxy derivatives. He failed to observe a rate-product correlation[42] of the kind that established π-participation in the 3-aryl-2-butyl system (Section 4.6). The absence of the rate-product correlation has led him to question the presence of σ-bridged cations in the solvolysis of 1-methyl-2-adamantyl tosylate.

14.6. Conclusion

Both the hyperconjugative model of Jensen[3] and Traylor[4] and the *exo*-6–H participation model of Olah[5] appear to require preferential electron supply from the *exo* direction. The data available at present do give indications of enhanced electron supply from the strained σ-bonds of the norbornyl system, but fail to reveal a significant preference for electron supply from the *exo* as compared to the *endo* direction.

A number of new systems have been proposed as involving the formation of $\pi\sigma$- or σ-bridged cations (Section 14.5). Of these, the most favorable case appears to be the Coates' cation (**44** or **45**).[32,33] However, it is evident that many proposals for such bridged cations are still being advanced without the definite evidence that is considered essential for structure assignments in other areas of organic chemistry.[43]

[40]F. T. Bond, H. L. Jones, and L. Scerbo, *Org. Photochem. Synth.*, **1**, 33 (1971).

[41]G. A. Olah, G. Liang, and S. P. Jindal, *J. Amer. Chem. Soc.*, **98**, 2508 (1976).

[42]D. Farcasiu, *J. Amer. Chem. Soc.*, **98**, 5301 (1976).

[43][Note added in proof.] Application of the tool of increasing electron demand to the 9-aryl-9-pentacyclo[4.3.0.0^{2,4}.0^{3,8}.0^{5,7}]nonyl system reveals $\rho+$ -2.05, as compared to -5.27 for the parent 7-aryl-7-norbornyl derivatives. Consequently, this tool establishes that the system undergoes solvolysis with major participation by the $\pi\sigma$-electrons of the cyclopropane ring and supports Coates' conclusion[32,33] that the intermediate is a carbon-bridged nonclassical ion.

Comments

I agree with Brown that the hyperconjugative model does not provide a satisfactory explanation for 2-norbornyl behavior. Although hyperconjugation and σ-electron participation are closely related phenomena, the former implies insignificant nuclear movement towards bridging. Without such nuclear movement, it is very difficult to rationalize the rate decreases produced by remote alkyl substitutions (24). (See Chapter 10 for data and further examples.)

$$\text{RR(25°C):} \qquad 1.0 \qquad\qquad 0.0138 \qquad\qquad 0.012 \tag{24}$$

Hyperconjugation effects diminish in magnitude with increasing carbocation stability.[44] If differential hyperconjugation were responsible for 2-norbornyl behavior, one would expect *exo*: *endo* ratios to *decrease* along the series: secondary, methyl-substituted tertiary, aryl-substituted tertiary, contrary to observation. For the same reason, it does not seem satisfactory to attribute the high *exo*: *endo* tertiary ratios to differential hyperconjugation.

Evidence indicating better electron donation by a 2-*exo*- than a 2-*endo*-norbornyl *substituent* (in 2-norbornyl mercurenium ion systems) has now appeared,[45] but I do not feel that the results are very relevant to the question of C1,6 participation.

The methyl hydrogen rate ratios summarized in Table 14.3 do indeed show fortuitously similar values for *exo*- and for *endo*-norbornyl, but this does not mean "that enhanced electron supply is occurring in both *exo* and *endo* isomers." The *exo*-methyl:hydrogen ratio is decreased because of participation during 2-*exo*-norbornyl trifluoroacetolysis; tertiary *exo* solvolysis is normal. Trifluoroacetolysis of 2-*endo*-norbornyl tosylate is normal, but steric inhibition toward ionization preferentially affects the tertiary *endo* derivative, and an abnormally low methyl:hydrogen ratio results.

Work completed within the last few months has now convinced Professor Brown that Coates' cation **44** is bridged.[46-48] This system is important because it provides an undisputed model which calibrates the behavior of

[44]L. Radom, J. A. Pople, and P. v. R. Schleyer, *J. Amer. Chem. Soc.*, **94**, 5935 (1972).
[45]W. A. Nugent, M. M.-H. Wu, T. P. Fehlner, and J. K. Kochi, *Chem. Commun.*, 456 (1976).
[46]W. L. Jorgensen, *Tetrahedron Lett.*, 3033 (1976).
[47]R. M. Coates and E. R. Fretz, *J. Amer. Chem. Soc.*, **99**, 297 (1977).
[48]H. C. Brown and M. Ravindranathan, *J. Amer. Chem. Soc.*, **99**, 299 (1977).

other nonclassical ions. A more detailed examination of the facts which led
to this conclusion is warranted.

	32	**43**
RR(25°C): 1.0	10^{11}	10^{10-12}

Coates' original evidence already was conclusive.[32,33,49] The rate
enhancement of **43**, relative to 7-norbornyl, is enormous; σ-participation in
43 is comparable to π-participation in **32**. The rate enhancement of **43**
cannot be attributed to the formation of a rearranged ion with a lower strain
content. In this system, the rearrangements are degenerate and only lead to
products with the same gross structure as the starting material. Label
scrambling (23) reveals an extent of rearrangement which rigorously excludes "a set of rapidly equilibrating classical ions or ion pairs" (**45a–45e**,
etc.).

45a	**45b**	**45c**	**45d**	**45e**
	⇅	⇅	⇅	⇅
	etc.	etc.	etc.	etc.

Because of the C_{2v} symmetry of the classical ion (**45a**), positions C2, C3, C6,
and C7 are all equivalent, and, in the first stage of a rapid equilibrium,
deuterium label originally on C9 should be scrambled among *five* and not
just *three* positions. But this is not all. Further rapid rearrangement from
ions **45b–45e** should scramble more positions, and such processes should
continue until all *nine* CH groups become equilibrated.[49]

It could be argued that ion pair phenomena account for the observed
result (23) or that the ions have insufficient time to scramble fully. But the
stable ion in superacid shows C_{3v} symmetry (**54** ≡ **44**); that is, chemical shifts
for *three different types* (a, b, and c) of ^{1}H and of ^{13}C atoms are observed.[33] If
the ion were rapidly equilibrating (**45a–45e**, etc.), only *one* ^{1}H and *one* ^{13}C
signal would have been expected under those conditions.

[49]Review: R. E. Leone, J. C. Barborak, and P. v. R. Schleyer, *Carbonium Ions*, Vol. IV, G. A.
Olah and P. v. R. Schleyer, Eds., Wiley–Interscience, New York, 1973, Chapter 33, pp.
1869–1872.

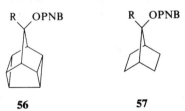

54   **55a**   **55b**

Coates and Fretz thus not only proved that **54** was a nonclassical ion, but they demonstrated a very high degree of structural stability as well. Thus, the degenerate "bridge flipping," **55a⇌55b**, was shown to have an unmeasurably large barrier for the parent (R = H), and a barrier of $\Delta G^{\ddagger}_{-8°} =$ 13.0 kcal/mole for the tertiary ion (R = CH$_3$). MINDO/3 calculations reveal C_{3v} symmetry for the parent ion (**54**), and estimate bridge flipping barriers (**55a⇌55b**, assuming C_{2v} symmetry for the transition state) of 22.1 kcal/mole for R = H and 8.8 kcal/mole for R = CH$_3$.[46]

The new experimental evidence confirms the trishomocyclopropenium nonclassical structure of this ion and its derivatives. Remarkably low solvolytic α-CH$_3$/H rate ratios (**56**, R = CH$_3$ and H) of 112[47] or 127,[48] and α-C$_6$H$_5$/H ratios (**56**, R = C$_6$H$_5$ and H) of 1600[47] or 1800[48] are found by the two independent investigations. Anchimeric assistance of large magnitude is present even in the *tertiary* derivatives, 10$^{7.5}$ for **56** (R = CH$_3$) and 10$^{3.6}$ (R = C$_6$H$_5$) relative to **57** (R = CH$_3$ or C$_6$H$_5$).[47] The "tool of increasing electron demand" applied to **56** (R = aryl) gives ρ+ −2.05.[48] Although no "break" was found in the Hammett–Brown plot, the ρ+ value is so low in absolute magnitude (compare ρ+ −5.27 for **57**, R = aryl, Chapter 10) that it seems difficult to find fault with Brown's and Ravindranathan's conclusion[48] "that the solvolysis of this system [**56**] proceeds with $\pi\sigma$-participation."

R   OPNB   R   OPNB

56   **57**

The nonclassical stonewall, once breached, should rapidly crumble.

In view of the detailed structural information available for analogous carboranes (see comments, Chapter 1), there would seem to be less need to "reserve judgment" about the correctness of the C_{5v} formulation **48** of Hogeveen's dication. The proposal of rapidly equilibrating alternatives (e.g., **51**) are an attempt to preserve the idea that carbon can only form two-center two-electron bonds. The need and the desirability of such preservation is long past. How can the many carboranes (Comments, Chapter 1), in which each carbon atom and each boron atom is penta- or hexacoordinate, be formulated in terms of rapidly equilibrating Lewis structures? Instead, we should be embracing species like **48** and **54** with enthusiasm for what they teach us about bonding, particularly since these "new concepts" are easy to understand from a theoretical viewpoint.

A system like **43**, with σ-participation of the order 10^{10}–10^{12}, was required to provide an undisputed example of a σ-bridged nonclassical ion. The structural variety of organic compounds is so great that a continuum of behavior is to be anticipated. It is almost certain that σ-participation of smaller magnitude, $k_\Delta/k_c = 10^8$, 10^5, 10^2, or even 10^1, has been observed in at least some of the many systems studied to date. Nothing argues against such a possibility; theory supports, rather than refutes, the existence of nonclassical carbocations. The only problem has been the *unambiguous* demonstration of bridging to everyone's satisfaction. Evidence is available in abundance, only the interpretation of the evidence is in question.

The 2-norbornyl, the 2-bicyclo[2.1.1]hexyl (**49**), and the 1-methyl-2-adamantyl (**50**) cations represent a few of the cases toward the lower end of the participation scale.

58

Anchimeric assistance in **58** ($R = CH_3$), leading to **50**, has been estimated as the basis of several lines of argument to be only on the order of 10–30.[36] The evidence is conceded to be *consistent* with the nonclassical interpretation; the dispute is whether or not the evidence *requires* such a formulation.[42]

Farcasiu's argument against bridging, cited by Brown, is in error. *No rate-product correlation is to be expected unless two competing, discrete mechanisms are involved.* This is the case in the 3-phenyl-2-butyl systems (Section 4.6) where solvent assistance, k_s, competes against anchimeric assistance, k_Δ. The *rate determining steps* for the k_s and for the k_Δ processes *also control the products*, because no crossover between pathways occurs. Both k_s and k_Δ contribute to the total rate constant, k_t ($k_t = k_s + k_\Delta$).

The 2-adamantyl system is fundamentally different. Because of the crowding involved, solvent assistance is not possible and $k_s \equiv k_c$, i.e., anchimerically unassisted solvolysis is limiting. The completely unassisted rate constant, k_c, represents a process which is not *discrete*, but is the limit to which both k_Δ and k_s tend in the absence of assistance. Thus, the equation,

$k_t = k_c + k_\Delta$, is *not* a proper one.[50] A classical ion can be formed in a rate limiting step (k_c) and rearrange subsequently. The amount of rearrangement would be controlled by the conditions, not by any inherent property of the carbocation system, and no rate-product correlation is expected. Likewise, a nonclassical ion like **59** gives a different product distribution in

different solvents or under different conditions, but the *magnitude* of anchimeric assistance remains very large $(\sim 10^{11})$ in all circumstances.[51] No rate-product correlation exists or is to be expected.

The experiment suggested by Farcasiu[42] has now been carried out. Solvolysis of optically active **58** (R = CH₃) in 50% aqueous acetone proceeds with 90% retention of configuration.[52] This result is consistent with a weakly bridged ion, **50**, and the modest magnitude of rate enhancement produced by methyl substitution.[36] The classical secondary ion alternative **53** is achiral, as it possesses a plane of symmetry, and does not seem to provide a very likely rationalization of these results.

[50]P. v. R. Schleyer, J. L. Fry, L. K. M. Lam, and C. J. Lancelot, *J. Amer. Chem. Soc.*, **92**, 2542 (1970), see especially footnote 5.
[51]A. Diaz, M. Brookhart, and S. Winstein, *J. Amer. Chem. Soc.*, **88**, 3133 (1966). Also, see Section 4.6.
[52]U. Göckel, Diplomarbeit, Universität Erlangen-Nürnberg (1977).

15

Final Comments

15.1. Introduction

It has long been recognized that saturated structures with electron-deficient cationic centers are highly reactive species, undergoing both rapid rearrangements and rapid reactions with external nucleophiles.[1,2] Such an electron-deficient cationic center will undertake to reduce its electron deficiency by internal electron shifts, delocalizing the charge throughout the carbon structure, or by interacting with the solvent or other nucleophiles in its vicinity.

In a solvolytic process, involving the generation of such a carbonium ion, such stabilization occurs either internally (inductive effects, field effects, hyperconjugation), or externally (solvent participation).

In the early 1950s it was proposed that there existed a major new factor which could stabilize such cations—σ-bridging between the cationic center and a neighboring carbon–carbon[3,4] or carbon–hydrogen bond.[5] It was proposed that this new factor greatly stabilized carbonium ions and could account for the enhanced rates of solvolysis exhibited by many highly branched organic derivatives, an enhanced reactivity previously attributed to the driving force provided by relief of steric strain.[6,7]

[1] H. Meerwein and K. van Emster, *Ber.*, **55**, 2500 (1922).
[2] F. C. Whitmore, *Chem. Eng. News*, **26**, 668 (1948).
[3] J. D. Roberts and R. H. Mazur, *J. Amer. Chem. Soc.*, **73**, 3542 (1951).
[4] S. Winstein and D. Trifan, *J. Amer. Chem. Soc.*, **74**, 1147, 1154 (1952).
[5] J. D. Roberts and J. A. Yancy, *J. Amer. Chem. Soc.*, **74**, 5943 (1952).
[6] H. C. Brown and R. S. Fletcher, *J. Amer. Chem. Soc.*, **71**, 1845 (1949).
[7] H. C. Brown and H. Berneis, *J. Amer. Chem. Soc.*, **75**, 10 (1953).

15.2. The Search for σ-Participation and σ-Bridging

Such σ-bridging was indeed utilized to account for the observed fast rates of solvolysis of camphene hydrochloride,[8] tri-*tert*-butylcarbinyl derivatives,[9] and cyclodecyl tosylate.[10] These developments stimulated a program to test for the presence of σ-bridging as a major factor in the chemistry of such carbonium ions. These proposals were soon disproved and withdrawn.[11-13]

Although my personal stake in the question thereby vanished, I had become personally interested in the general question of whether such σ-bridged cations exist. It was generally agreed that cyclopropylcarbinyl[3] and 2-norbornyl provided the best examples for σ-bridging.[14] It proved to be a most difficult task to arrive at a definitive answer.

In this book we have presented the evidence and arguments pro and con σ-bridged structures for these cations. As far as I can see, the evidence does not support the presence of such σ-bridges in these cations. The reader may, however, arrive at the opposite conclusion. This illustrates the difficulties one can encounter in deciding between alternative interpretations for a given set of observations. This situation is rarely presented to students. Yet it is important for them to recognize the existence of this problem.

15.3. Does σ-Delocalization Exist?

It should be recognized that the problem does not involve the question of the existence of σ-delocalization in cations. Hyperconjugation was proposed as early as 1941[15] and incorporated into textbooks as early as 1943 to account for the increasing order of stability of carbonium ions, $Me^+ < Et^+ < i\text{-}Pr^+ < t\text{-}Bu^+$.[16]

The proposal in early 1950 was that in such cations, superimposed upon the usual delocalization attributed to inductive, field, and hyperconjugative effects, there was a new phenomenon, σ-bridging, which could not only stabilize the system over and above the stabilization provided by the earlier

[8]F. Brown, E. D. Hughes, C. K. Ingold, and J. F. Smith, *Nature*, **168**, 65 (1951).
[9]P. D. Bartlett, *J. Chem. Educ.*, **30**, 22 (1953).
[10]R. Heck and V. Prelog, *Helv. Chim. Acta*, **38**, 1541 (1955).
[11]H. C. Brown and F. J. Chloupek, *J. Amer. Chem. Soc.*, **85**, 2322 (1963).
[12]P. D. Bartlett and T. T. Tidwell, *J. Amer. Chem. Soc.*, **90**, 4421 (1968).
[13]V. Prelog, *Rec. Chem. Progr.*, **18**, 247 (1957).
[14]P. D. Bartlett, *Nonclassical Ions*, Benjamin, New York (1965).
[15]R. S. Mulliken, C. A. Rieke, and W. G. Brown, *J. Amer. Chem. Soc.*, **63**, 41 (1941).
[16]A. E. Remick, *Electronic Interpretations of Organic Chemistry*, Wiley, New York, 1943.

effects, but could also alter the symmetry properties of the cations in a major way.

Thus the 2-norbornyl cation, an unsymmetrical species in its classical form, was assigned a symmetrical structure in its nonclassical form.[4] Indeed, it retains that symmetry in the many alternative nonclassical structures which have been proposed (Figure 6.1). Only Traylor in his hyperconjugative structure (Section 14.2) terms "nonclassical" an unsymmetrical structure, essentially identical with the classical structure for the 2-norbornyl cation.

It is here urged that the term "nonclassical" be reserved for σ-bridged cations in accordance with the definition proposed (Section 4.2).

15.4. Are There σ-Bridged Cations?

In the early 1950s the enthusiasm for the new concept was so great that σ-bridged structures were at least considered for every known cation (i.e., ethyl, *n*-propyl, 2-butyl, 2,3,3-trimethyl-2-butyl, etc.) with the possible exception of methyl. The situation at that time can be represented graphically as shown in Figure 15.1.

Over the years essentially all of the cations originally explored as possibilities for σ-bridged structures have been reclassified as classical ions. Even under stable ion conditions, the great majority of ions are now considered to be either static classical ions or rapidly equilibrating cations (Table 13.1). Consequently, the present situation with respect to these cations can be represented graphically as shown in Figure 15.1.

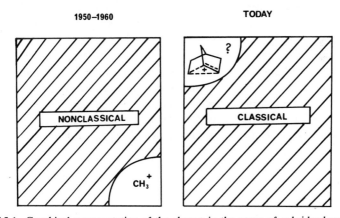

Figure 15.1. Graphical representation of the change in the scope of σ-bridged nonclassical ions.

In Section 14.5 there were presented a number of new systems where σ-bridged intermediates have been proposed. Some of these look promising, but the definitive experiments have yet to be carried out. Should these prove to have σ-bridged structures, Figure 15.1 would require only minor revision. It is quite clear that the huge majority of solvated cations normally encountered in organic reactions have classical structures without σ-bridges. σ-Bridged cations, if they are confirmed, would appear to be significant only in a relatively narrow, highly specialized group of structures, of theoretical interest, but of little practical significance. They could be accommodated in the corner now reserved for 2-norbornyl (?).

15.5. Position on σ-Bridged Cations

Before completing this review of the existence of σ-bridged cations, it may be appropriate to clear up some misconceptions of my position.

Why do I reject σ-bridged cations, when three center–two electron bonding is evident in diborane and dimeric trimethylaluminum, molecules with which I am surely not unfamiliar?

My questioning of the nonclassical ion concept is not based upon any theoretical considerations. It is based entirely on the repeated failure in our studies to find convincing evidence to support σ-bridged structures in many systems where such structures had been proposed for species produced in solvolytic processes.

As was discussed much earlier,[17] molecules such as diborane and trimethylaluminum dimer, exist in the dimeric form in the gas phase and in nonpolar solvents. Diborane in solvents such as tetrahydrofuran, and trimethylaluminum dimer, in solvents such as ethyl ether, exist in monomeric form, actually as the complexes, $H_3B:O(CH_2)_4$ and $(CH_3)_3Al:O(C_2H_5)_2$. We are interested in the structure of carbonium ions produced and transformed in solvolytic processes or in other carbonium ion reactions. Consequently, it does not appear appropriate to assume the existence of such σ-bridged cationic structures in the absence of convincing experimental data requiring such structures.

Do I deny the possibility of the existence of σ-bridged cations? Again the answer is already in print, if critics would only read my writings on the subject.[18]

To quote:[18] "I have found it difficult to recognize my own position from that attributed to me in some recent publications [G. D. Sargent, *Quart. Rev.*

[17]H. C. Brown, "The Transition State," *Chem. Soc.* (*London*), *Spec. Publ.*, **16**, pp. 140–158, 174–178 (1962).

[18]H. C. Brown, *Chem. Eng. News*, **45**, 86, Feb. 13, 1967.

(*London*), **20**, 301 (1966)]. For example, in a recent book review of *Nonclassical Ions*, it was stated [R. J. P. Williams, *Nature*, 455 (1966)]: '. . . I also found a short review in *Chemistry in Britain* (199, 1966) which denies the whole concept. I do not advise my fellow students to use this book unless they want a nonclassical outlook on neoclassical ions.' As the author of the short review referred to by Williams, I wish to make it clear that in my review I did not deny the whole concept. (It would be unscientific to take a dogmatic position that any particular phenomenon is incapable of existence in this fascinating, versatile world of ours.) My position then and now is merely that my students and I have been unable to find any experimental evidence whatever for σ-bridging in the solvolysis of norbornyl derivatives."

Do I deny the existence of three center–two electron bonds? Of course not. They have been established in diborane and trimethylaluminum dimer. Their general occurrence in solvated carbonium ions is yet to be demonstrated. Many, many structures for carbonium ions containing σ-bridges of this kind have been advanced. Essentially all of these, which have been subjected to critical examination, have disappeared. In view of this history, it would appear desirable that those who now propose carbonium ion structures containing such σ-bridges (Section 14.5) should support such proposals with definite unambiguous experimental evidence.

15.6. Conclusion

We have come a long way since a particular nonclassical structure for a carbonium ion was proposed "Because it looks so nice," or "Because it is fashionable," or "Only a dry, undistillable resin remained upon removal of the solvent. This suggests that a nonclassical ion is involved in the mechanism."[19] It is to be hoped that we will heed the counsel of R. B. Woodward that the mere fact that one is dealing with fugitive intermediate should not convey license to propose highly fanciful structures without adequate experimental support.[20]

Another eminent chemist, Joseph Priestley, involved in an earlier controversy concerning phlogiston, warned us that "a philosopher who has been long attached to a favorite hypothesis, and especially if he has distinguished himself by his ingenuity in discovering or pursuing it, will not sometimes be convinced of its falsity by the plainest evidence of fact," and thus "both himself and all his followers are put upon false pursuits, and seem

[19] *Chem. Abstr.*, **60**, 9121b (1964).
[20] R. B. Woodward, *Perspective in Organic Chemistry*, A. Todd, Ed., Interscience, New York, 1956, pp. 177–178.

determined to warp the whole course of nature to suit their manner of conceiving of its operations."[21]

To retain perspective, it remains to be established whether it is the original proposal of σ-bridged cations or my challenging of the concept as a common general phenomenon in carbonium ion chemistry which constitutes the "phlogiston theory" of the current century!

This has been a fascinating research project with numerous unexpected turns in the research trail. A careful consideration of the arguments and evidence should provide a valuable experience for the student or other reader. A major disappointment to the author has been the intense emotion that his questioning of an apparently established concept aroused. It proved impossible to achieve an open discussion of the evidence and arguments. As mentioned earlier, various editors found it impossible to present a "symposium-in-print" on the subject. The present publication with the commentary by Paul v. R. Schleyer is the first time that both points of view have been presented together for convenient consideration of the available evidence and alternative interpretations. One can only hope that the next generation will find it less difficult to discuss and resolve this and similar differences in position and interpretation objectively, scientifically, and with a sense of humor.

[21]P. Hartog, *Memorial Lectures, The Chemical Society (London)*, **4**, 21 (1951).

Comments

Since I have devoted most of my comments to criticisms of Brown's evidence, interpretations, and positions, let me end more positively by acknowledging his contributions to this area. Much, besides the structures of a few charged species, is to be learned from these fifteen years of controversy.

The passage of time affords perspective. It is instructive to reread Brown's 1962 article,[22] his first publication on "the question of nonclassical carbonium ions." It is difficult now to find very much objectionable in it. Some of Brown's criticisms (e.g., cyclodecyl) have proved to be valid; others (e.g., the phenonium ion) did not. The rapidly equilibrating set of bicyclobutonium ions is no longer with us. He raised legitimate questions that required answers. When answers were found, our understanding benefited. For example, Brown's stress on the lack of rate enhancement due to β-phenyl substitution in 2-butyl systems led directly to our present appreciation of the role of solvent assistance in solvolysis.

In retrospect, it does not detract from the achievements of the time to say that we knew relatively little about carbocations in 1962. Direct studies in superacid media had just commenced.[23] Refined quantum mechanical calculations had not been applied.[24] The numerous physical tools now available (nmr, ESCA, icr, etc.) had not been utilized.[23] Were the commonly accepted interpretations of 1962 really on unassailably secure ground? We think we are in a much better position today, but what will the state of our understanding be in 1992?

Major problems remain to be solved. No general method has yet been developed for the experimental determination of detailed structures of carbocations in the gas phase. No direct experimental information concerning the structure of CH_3^+ exists! Even X-ray structures of stable carbocations have been confined to a few highly resonance stabilized species.[25] The energies of very few carbocations have been determined by physical means. Solvation is poorly understood. We are not yet able to predict solvolysis rates accurately for a wide variety of systems.

Major developments will continue. Even as this book was being written, the existence of an unambiguous nonclassical carbocation was established. This marks the beginning of the end of this controversy, but the need for careful scientific work, soundly based interpretations, and constructive criticism will always remain.

[22]See Ref. 19, Chap. 1.

[23]For a review, see G. A. Olah, *Carbocations and Electrophilic Reactions*, Wiley, New York, 1974.

[24]Review: L. Radom, D. Poppinger, and R. C. Haddon, *Carbonium Ions*, Vol. V, Chap. 38, pp. 2303ff, G. A. Olah and P. v. R. Schleyer, Eds., Wiley–Interscience, New York, 1976.

[25]Review: M. Sundaralingam and A. K. Chwang, *Carbonium Ions*, Vol. V, Chap. 39, pp. 2427ff, G. A. Olah and P. v. R. Schleyer, Eds., Wiley–Interscience, New York, 1976.

Epilog

The author cannot resist the prerogative afforded by his position to present the final word. σ-Bridged nonclassical ions were originally proposed to account for fast rates and unusual stereochemistry in solvolytic processes involving carbocationic intermediates. As is evident from the discussion in this book, none of the structures originally proposed and considered are still with us. Indeed, of the two examples believed originally to provide the most convincing cases for such σ-bridged carbocationic intermediates, cyclopropylcarbinyl and 2-norbornyl, Professor Paul Schleyer in his Comments rules out the bicyclobutonium formulation for the former, and in a letter to the author (1975) termed the latter, the 2-norbornyl cation, to be "ambiguous."

The discussion currently appears to be receding from the original problem, the structure of carbocations under solvolytic conditions. Instead, current studies emphasize the structure of carbocations under stable ion conditions. More recently, the emphasis has been directed toward the structure of carbocations in the gas phase.

In superacid media, equilibrating classical carbocations appear to have finally come into their own (Table 13.1), with only cyclopropylcarbinyl and 2-norbornyl, of the original systems considered, still assigned σ-bridged structures by some workers in the field. The gas-phase studies have hitherto ignored the probable effect of solvation in stabilizing the more localized charges of the classical structures over the more delocalized charges of the corresponding nonclassical structures.

A further development has been the increasing complexity of the systems now being proposed as examples of σ-bridged structures. We have progressed from the simple aliphatic and alicyclic cations originally considered, to the more complex cyclopropylcarbinyl and 2-norbornyl species, to the far more complex structures currently proposed (see

Comments 13). These complex structures are of considerable interest theoretically, but would appear to be of little significance in interpreting the behavior of the simpler carbocations that the organic chemist normally utilizes as intermediate in his reactions.

Finally, the author wishes to express his appreciation to Professor Paul Schleyer for the objectivity of his comments. It is to be hoped that others working in this fascinating area will exhibit equal scientific objectivity in presenting their own results and conclusions.

Subject Index